PATHOPHYSIOLOGICAL EFFECTS OF ENDOTOXINS AT THE CELLULAR LEVEL

PROGRESS IN CLINICAL AND BIOLOGICAL RESEARCH

RECENT TITLES

Vol 50: **Rights and Responsibilities in Modern Medicine: The Second Volume in a Series on Ethics, Humanism, and Medicine,** Marc D. Basson, *Editor*

Vol 51: **The Function of Red Blood Cells: Erythrocyte Pathobiology,** Donald F. H. Wallach, *Editor*

Vol 52: **Conduction Velocity Distributions: A Population Approach to Electrophysiology of Nerve,** Leslie J. Dorfman, Kenneth L. Cummins, and Larry J. Leifer, *Editors*

Vol 53: **Cancer Among Black Populations,** Curtis Mettlin and Gerald P. Murphy, *Editors*

Vol 54: **Connective Tissue Research: Chemistry, Biology, and Physiology,** Zdenek Deyl and Milan Adam, *Editors*

Vol 55: **The Red Cell: Fifth Ann Arbor Conference,** George J. Brewer, *Editor*

Vol 56: **Erythrocyte Membranes 2: Recent Clinical and Experimental Advances,** Walter C. Kruckeberg, John W. Eaton, and George J. Brewer, *Editors*

Vol 57: **Progress in Cancer Control,** Curtis Mettlin and Gerald P. Murphy, *Editors*

Vol 58: **The Lymphocyte,** Kenneth W. Sell and William V. Miller, *Editors*

Vol 59: **Eleventh International Congress of Anatomy,** Enrique Acosta Vidrio, *Editor-in-Chief.* Published in 3 Volumes:
Part A: **Glial and Neuronal Cell Biology,** Sergey Fedoroff, *Editor*
Part B: **Advances in the Morphology of Cells and Tissues,** Miguel A. Galina, *Editor*
Part C: **Biological Rhythms in Structure and Function,** Heinz von Mayersbach, Lawrence E. Scheving, and John E. Pauly, *Editors*

Vol 60: **Advances in Hemoglobin Analysis,** Samir M. Hanash and George J. Brewer, *Editors*

Vol 61: **Nutrition and Child Health: Perspectives for the 1980s,** Reginald C. Tsang and Buford Lee Nichols, Jr., *Editors*

Vol 62: **Pathophysiological Effects of Endotoxins at the Cellular Level,** Jeannine A. Majde and Robert J. Person, *Editors*

Vol 63: **Membrane Transport and Neuroreceptors,** Dale Oxender, Arthur Blume, Ivan Diamond, and C. Fred Fox, *Editors*

Vol 64: **Bacteriophage Assembly,** Michael S. DuBow, *Editor*

Vol 65: **Apheresis: Development, Applications, and Collection Procedures,** C. Harold Mielke, Jr., *Editor*

See pages 191–192 for previous titles in this series.

PATHOPHYSIOLOGICAL EFFECTS OF ENDOTOXINS AT THE CELLULAR LEVEL

Based on Proceedings of a Symposium
Sponsored by the Office of Naval Research
Held in Washington, DC

Editors
JEANNINE A. MAJDE
Office of Naval Research
Chicago Branch

ROBERT J. PERSON
University of Oklahoma
Health Sciences Center

ALAN R. LISS, INC., NEW YORK

Address all Inquiries to the Publisher
Alan R. Liss, Inc., 150 Fifth Avenue, New York, NY 10011

Printed in the United States of America.

Library of Congress Cataloging in Publication Data
Main entry under title:

Pathophysiological effects of endotoxins at the
cellular level.

(Progress in clinical and biological research;
62)

Includes index.
1. Endotoxin—Physiological effect—Congresses.
2. Endotoxin—Toxicology—Congresses. 3. Pathology, Cellular—Congresses. I. Majde, Jeannine A.
II. Person, Robert J. (Robert James) III. United
States. Office of Naval Research. IV. Series.
[DNLM: 1. Endotoxins—Pharmacodynamics—Congresses.
W1 PR668E v.62 / QW 630 P297 1979]
QP632.E4P37 615.9 81-6066
ISBN 0-8451-0062-9 AACR2

Contents

Contributors

S. Gaylen Bradley
Departments of Microbiology and Pharmacology, P.O. Box 678, Virginia Commonwealth University, Richmond, VA 23298 **[3]**

Delores Di Pietro
Department of Microbiology, Emory University School of Medicine, Atlanta, GA 30322 **[47]**

J. Raymond Fletcher, USN
Experimental Surgery Division, Naval Medical Research Institute, National Naval Medical Center, Bethesda, MD 20014 **[173]**

Michael L. Hess
Department of Medicine (Cardiology), Medical College of Virginia, P.O. Box 281, MCV Station, Richmond, VA 23298 **[23]**

Stephen M. Krause
Department of Physiology, Medical College of Virginia, P.O. Box 281, MCV Station, Richmond, VA 23298 **[23]**

Glenn J. Leach
Department of Physiology, Louisiana State University Medical Center, 1542 Tulane Avenue, New Orleans, LA 70112 **[123]**

Maw-Shung Liu
Department of Physiology and Pharmacology, Bowman Gray School of Medicine, Wake Forest University, Winston-Salem, NC 27103 **[115]**

Will M. Long
Department of Physiology, Louisiana State University Medical Center, 1542 Tulane Avenue, New Orleans, LA 70112 **[115]**

Ronald V. Maier
Department of Immunopathology, Scripps Clinic and Research Foundation, 10666 North Torrey Pines Road, La Jolla, CA 92037 **[133]**

Jeannine A. Majde
Office of Naval Research, Chicago Branch, 536 South Clark Street, Chicago, IL 60605 **[xi]**

John C. Mathison
Department of Immunopathology, Scripps Clinic and Research Foundation, 10666 North Torrey Pines Road, La Jolla, CA 92037 **[133]**

The number in bold type following the contributor's affiliation is the first page of that contributor's article.

Roderick E. McCallum
Department of Microbiology and Immunology, University of Oklahoma Health Sciences Center, P.O. Box 26901, Oklahoma City, OK 73190 **[99]**

Leena Mela
Department of Physiology and Surgery, Michigan State University, East Lansing, MI 48824 **[15]**

Robert J. Moon
Department of Microbiology and Public Health, Michigan State University, East Lansing, MI 48824 **[33]**

David C. Morrison
Department of Microbiology, Emory University School of Medicine, Atlanta, GA 30322 **[xi, 47, 157]**

Priscilla Munkenbeck
Department of Microbiology, Emory University School of Medicine, Atlanta, GA 30322 **[157]**

Zenaida G. Oades
Department of Microbiology, Emory University School of Medicine, Atlanta, GA 30322 **[47]**

Robert J. Person
Department of Physiology and Biophysics, University of Oklahoma Health Sciences Center, P.O. Box 26901, Oklahoma City, OK 73190 **[65]**

James A. Reeves
Department of Physiology, Louisiana State University Medical Center, 1542 Tulane Avenue, New Orleans, LA 70112 **[123]**

Ian A. Silver
Department of Pathology, University of Bristol Medical School, University Walk, Bristol 1TD, England **[81]**

John J. Spitzer
Department of Physiology, Louisiana State University Medical Center, 1542 Tulane Avenue, New Orleans, LA 70112 **[115]**

Judy A. Spitzer
Department of Physiology, Louisiana State University Medical Center, 1542 Tulane Avenue, New Orleans, LA 70112 **[123]**

Hassan Tavakoli
Department of Microbiology and Public Health, Michigan State University, East Lansing, MI 48824 **[33]**

Richard J. Ulevitch
Department of Immunopathology, Scripps Clinic and Research Foundation, 10666 North Torrey Pines Road, La Jolla, CA 92037 **[133]**

Richard I. Walker, USN
Medical Microbiology Division, Naval Medical Research Institute, National Naval Medical Center, Bethesda, MD 20014 **[173]**

Mark E. Wilson
Department of Microbiology, Emory University School of Medicine, Atlanta, GA 30322 **[157]**

Preface

Infection with Gram-negative bacteria continues to be a major clinical problem both in the United States and throughout the world. Recent studies have estimated 300,000 such cases in the United States annually with an approximate mortality of greater than 30 percent. Severe irreversible hypotensive shock, often accompanied by disseminated intravascular coagulation, contributes significantly to this mortality. Endotoxins have been implicated as major etiologic agents responsible for these pathophysiologic manifestations of Gram-negative sepsis. Evidence in support of this conclusion has derived from numerous studies in experimental animals where isolated purified endotoxins have been able to reproduce virtually all of the clinical responses, including shock and DIC, observed in patients with Gram-negative infections.

Intensive investigations during the past several decades have documented the importance of the lipopolysaccharide (LPS) component in inducing these pathophysiologic responses. These observations have been supported by experiments using protein-free LPS, LPS derived from polysaccharide deficient Gram-negative bacteria, and isolated purified lipid A. The administration of such endotoxins to mice, rats, rabbits, dogs, monkeys and sheep has provided extensive valuable information relevant to the biological activities of endotoxins and their potential deleterious role(s) in Gram-negative infections.

In parallel with these in vivo studies have been equally important studies in vitro which have demonstrated an impressive array of host defense systems which recognize the presence of bacterial endotoxins and, in many cases, become activated by these macromolecules. These include the non-specific host mediation systems as well as specific immunologic systems. What has become apparent to many investigators is that endotoxins have the capacity to interact with virtually every host cellular (and humoral) system with which they come into contact.

The realization of this conclusion, however, also suggests that efforts to define those interactions between endotoxins and host cells and tissues which may be of critical importance in patients with hypotensive shock resulting from Gram-negative sepsis may be extremely difficult. Thus, even today, pathogenesis of Gram-negative shock and DIC remains as an area of intensive experimental investigation. It is, however, becoming increasingly clear that the major factors which regulate the endotoxic activities of these potent bacterial products are their capacity to interact with the various host cells and tissues. Put simply, it is not only which tissues endotoxins localize to following their release from a Gram-negative organism, but rather what they do once they get there. Recent studies on genetically defined endotoxin-resistant inbred animals have, in fact, provided direct experimental support for this concept.

It is clear that it is the relative capacity of the lipid A region of endotoxins to interact with cells and tissues which regulates its endotoxic activity. It is not unreasonable, therefore, to suggest that there must be biochemical and/or biophysical properties unique to the lipid A which endow this molecule with its almost ubiquitous capacity to perturb mammalian cell membranes. It is interesting to speculate on the potential role of these nonspecific perturbational events as opposed to specific endotoxin receptors as contributing to this ability of endotoxins to affect mammalian cells.

It is nevertheless striking that endotoxins can and do perturb a wide variety of cells with diverse biological activities and functions. While investigators have recognized for many years the potential capacity for endotoxins to interact with lymphoid and reticuloendothelial cells, recent studies have expanded significantly the spectrum of cellular targets for endotoxic activity. It is thus apparent that a meaningful analysis of in vivo pathophysiological responses to endotoxins must include potential considerations of direct interactions with all major cell types of the body. Only through such investigations can we gain an understanding of events leading to release of secondary mediators, and more importantly, design rational approaches to therapy.

It was with the objective of defining further the "unity of diversity" of effects of endotoxins on mammalian cells that the Symposium on the Pathophysiology of Endotoxins at the Cellular Level was organized by the Office of Naval Research in October of 1979. This symposium brought together investigators with varied and unique areas of expertise, but each of whom shared a basic interest in the direct effects of endotoxins on mammalian cells. Unexpected parallels in the in vitro elicitation of endotoxin-induced responses, as well as some rather notable exceptions, were reported. This monograph summarizes the information presented at this symposium*, supplemented with three invited papers which could not be fitted into the program at the time. It documents the diversity of cellular and sub-cellular targets of endotoxins, and points to molecular events, such as calcium ion flux alterations by endotoxins (cf papers by Mela, Krause and Hess, Morrison et al., Person, J.A. Spitzer et al.), which suggest promising new therapeutic approaches in sepsis.

It is our hope that this monograph will contribute to interdisciplinary collaborative efforts between investigators and clinicians with the objective of the eventual elucidation of the molecular mechanisms of endotoxic interactions at cell membranes and their contribution to the pathogenesis of Gram-negative shock and DIC.

David C. Morrison
Jeannine A. Majde

*The papers collected here were prepared after the symposium; data and references are current as of mid-1980.

Editors' Note

Endotoxins are high molecular weight, biochemically complex macromolecules whose chemical composition is known to vary depending upon the method used for extraction, and whose physical form may depend critically upon the medium in which they are suspended. Thus the form of endotoxin selected for study may significantly affect the outcome of the experiment. This may be more of a problem when working in in vitro than in in vivo studies, as the culture media used lack the enzymes and carrier proteins found in vivo (cf Maier, et al.) which produce "physiologic" endotoxins. Further, it is likely that the loss of protein which occurs during the Westphal phenol-water extraction process (but is retained during the Boivin trichloroacetic acid extraction process) significantly alters the interaction of the endotoxin molecule with a cell membrane. Finally, the change in sugar composition which occurs in endotoxins from different strains of bacteria may have subtle effects not yet recognized. For these reasons we have asked all contributors to define the endotoxin(s) used in their work as carefully as possible. We are pleased that several contributors have chosen to compare different forms of endotoxins (cf Mela, Morrison et al., Person, McCallum, Maier et al., Wilson et al., Walker et al.) and hope that this trend will continue.

The Editors

Subcellular Effects of Endotoxins

Direct Action of Bacterial Endotoxin on Cells, Mitochondria, and Lysosomes

S. Gaylen Bradley

Pathophysiology of Endotoxicosis

Virtually every organ is affected in an animal administered life-threatening doses of endotoxin [1]. Kidneys from mice receiving a low dose (1–2 mg/kg) or a high dose (10–20 mg/kg) of endotoxin are enlarged 18 hr after treatment, but not at 3 hr [2]. Micropathological examination of kidneys from mice treated with a low dose of endotoxin discloses that the glomeruli are relatively normal, but those of mice treated with a high dose of endotoxin are abnormal. There is interstitial edema in and around the glomeruli and a marked widening of Bowman's space. Leukocytic infiltration into the interstitial tissue is observed, and the vascular channels appear congested. Some hemolyzed erythrocytes are seen within the capillary tuft [3]. A low dose of endotoxin increases the blood urea nitrogen (BUN) concentration by threefold, and a high dose of endotoxin increases the BUN by fivefold after 18 hr [2].

Endotoxin elicits a granulocytosis by 6 hr, which increases until 24 hr, but subsides by 72 hr [4]. The hemoglobin level and hematocrit values of mice given a high dose, but not a low dose of endotoxin, are lower after 18 hr than those of normal mice. Leukocyte counts are the same in treated and untreated mice, but platelet counts are markedly decreased with a low or high dose of endotoxin. Fibrinogen levels are elevated by a low dose of endotoxin but are dramatically reduced with a high dose of endotoxin at 18 hr. Prothrombin activity is not altered by a low dose of endotoxin but is severely decreased in mice given a high dose of endotoxin [3]. The concentration of cyclic adenosine monophosphate (cyclic AMP) in plasma of endotoxin-treated mice is increased after 3, 6, and 18 hr [2].

Liver weight and the concentration of cyclic AMP in mouse liver are not significantly altered by a low dose or a high dose of endotoxin [2]. Mice treated

Pathophysiological Effects of Endotoxins at the Cellular Level, pages 3–14

with a lethal dose of endotoxin show elevated serum glutamate–pyruvate transaminase (SGPT) activity after 1, 4, and 24 hr. With a low dose of endotoxin, SGPT activity is elevated at 1 and 4 hr, but has returned to normal values by 24 hr. Sulfobromophthalein (SBP) clearance is prolonged in mice treated with a low dose or a high dose of endotoxin after 4, 6, and 18 hr. After 18 hr, however, SBP clearance in mice given a low dose of endotoxin is approaching the values characteristic of untreated mice [3]. Endotoxin markedly impairs the activities of hepatic mixed function oxidases, as measured by the prolongation of hexobarbital-induced sedation. Mice previously treated with phenobarbital to stimulate the activities of the hepatic mixed function oxidases are as susceptible to endotoxin as normal mice [5]. These studies on the effects of endotoxin on mice confirm and/or establish that (a) the critical events in endotoxicosis occur within 4 hr, and (b) changes in liver cells occur rapidly after endotoxin is administered [1].

Cellular Perturbations

Endotoxin elicits a variety of alterations in cellular activities and organellar function. Lysosomal enzyme activities are elevated in human embryonic lung (WI-38) cells in culture and in African green monkey kidney (Vero) cells treated with 10 μg of endotoxin/ml [6, 7]. The increase in lysosomal enzyme activities is progressive with time, achieving a twofold increase within 4 hr. The rise is inhibited by cycloheximide or actinomycin D, is temperature dependent, and occurs whether serum is present or absent. Because the changes in lysosomal enzyme activities occur in the absence of added serum, the role of antibody, complement, hormones, and lymphokines is minimized. Endotoxin does not enhance the level of lysosomal enzyme activities when added directly to the granular fraction.

Bacterial endotoxin elicits enhanced lysosomal enzyme activities in established normal cell cultures and primary cultures of spleen cells or peritoneal exudate cells from normal mice [8, 9], but not in several transformed cell lines [7] or primary culture of spleen cells from endotoxin-tolerant mice or genetically resistant C3H/HeJ mice [8]. The proposition that lysosomes may be the target for the effects of endotoxin is supported by the observation that lysosomes from endotoxin-treated mice are more fragile than those isolated from untreated mice [2].

The activities of malate dehydrogenase, succinate dehydrogenase, and adenylate kinase are decreased in the granular fraction of Vero cells exposed to 10 μg of endotoxin/ml for 2 hr. These changes in mitochondrial enzyme activities precede those for lysosomal enzymes and are elicited by lower amounts of endotoxin [7]. Clearly endotoxin acts directly on normal established cell cultures causing early changes in mitochondria under conditions in which the activities of catalase and hexokinase are not perturbed. Mitochondrial enzyme activities leak extensively from the isolated granular fraction exposed to endotoxin for 45 min. Accelerated loss of mitochondrial enzyme activities from the granular fraction to the soluble fraction is dependent upon the endotoxin concentration and incubation time [7]. Similar results have been obtained with primary cultures of mouse liver cells and subcellular fractions of mouse liver cells [10]. The activities of mitochondrial enzymes are decreased in the granular fraction of mouse liver cells exposed to endotoxin, and the mitochondrial enzymes leak into the supernatant fraction within 45 min after exposure to endotoxin. Loss of mitochondrial enzyme activity from the isolated granular fraction is accelerated by free lipid A as well as by diverse endotoxin preparations. Endotoxin causes isolated mitochondria from mouse liver to swell [10]. Endotoxin affects enzymes located in the inner membrane of the mitochondrion, the mitochondrial matrix, and the intermembrane space simultaneously. These results indicate that (a) lipid A is the toxic moiety of endotoxin, (b) endotoxin affects lysosomes, and (c) endotoxin disrupts the integrity of the mitochondrion.

Mitochondrial Injury

Endotoxin, even at 100 μg/mg of mitochondrial protein, does not impair reduction of the cytochromes when reduced nicotinamide adenine dinucleotide (NADH) or glutamate is the electron donor. Accordingly, endotoxin does not directly inhibit the electron carriers of the respiratory chain. Endotoxin at 100 μg/mg of mitochondrial protein impairs reduction of the cytochromes when succinate is the electron donor, indicating that endotoxin can act at, or prior to, the linkage of flavin adenine dinucleotide–iron sulfur protein to coenzyme Q-cytochrome $b \rightarrow$ cytochrome c_1 cytochrome c [10].

Endotoxin inhibits oxygen consumption by isolated mouse liver mitochondria when glutamate + malate is the substrate and adenosine-5′-diphosphate (ADP) is added (state 3 respiration) but has little effect when ADP is not added (state 4 respiration). Uncoupling does not seem to be the mechanism of action of endotoxin when glutamate is the substrate. Both endotoxin and lipid A cause a loss of respiratory control but the ADP/O ratios in isolated mouse liver mitochondria treated with endotoxin or lipid A are not altered significantly [10]. Impaired coupling of oxidative phosphorylation by endotoxin, therefore, is secondary to other interactions between endotoxin and a mitochondrial constituent.

Endotoxin or lipid A stimulates state 4 respiration by isolated mouse liver mitochondria when succinate is the substrate, but has little effect on state 3 respiration [11]. Accordingly, the manner by which endotoxin alters the respiratory capabilities of isolated mitochondria differs with various substrates. Endotoxin appears to act as an uncoupling agent when succinate is the substrate, in that state 4 respiration is stimulated. The amount of ADP utilized to the amount of O_2 consumed (ADP/O), however, is not markedly altered by endotoxin with either glutamate or succinate as the substrate, indicating that the coupling sites are functioning. In respiring mitochondria, the rate of oxygen consumption is determined by the concentration of ADP rather than by the concentration of the substrate. The dependence of the respiratory rate upon ADP, expressed as the RCR, is a useful measure of the intactness of isolated mitochondria. The deleterious effect of endotoxin or lipid A on respiratory control with either glutamate or succinate as the substrate may reflect damage to the integrity of the mitochondria rather than an effect on a particular site in electron transport or oxidative phosphorylation.

Current Investigations

There are important gaps in our knowledge about the action of endotoxin. Selected chapters in this monograph address a number of these: (a) Is endotoxin internalized, (b) what is the basis of endotoxin resistance in the C3H/HeJ mouse, and (c) what is the mechanism of endotoxin damage to mitochondria? In order to assess the specificity and biological significance of observed interactions involving endotoxin, it is necessary to relate effects on established cell cultures, primary cell cultures, subcellular fractions, and intact animals. Primary cell cultures and subcellular fractions may be prepared from normal mice, from mice rendered resistant to endotoxin by prior treatment with endotoxin, or from the genetically endotoxin-refractory C3H/HeJ mouse. In this study, the effects of endotoxin on Beige mice, which have impaired lysosomal function [12], have been examined in order to assess whether lysosomes are essential for the molecular and pathophysiological damage observed during endotoxicosis.

MATERIALS AND METHODS

Adult male C3H/HeDub mice weighing 18–22 g were obtained from Flow Research Animals Inc., Dublin, VA, and comparable male Beige mice were obtained from Jackson Laboratories, Bar Harbor, ME. Primary cell cultures were prepared from perfused mouse liver as described previously [10]. The cells were harvested, disrupted, and fractionated into a mitochondrial fraction (10,000g sediment), lysosomal fraction (20,000g sediment), and soluble fraction as described previously [7]. The specific activity of hexokinase was expressed as nanomoles of NADP reduced per minute per milligram of protein, β-glucuronidase activity as micrograms of phenolphthalein released per minute per milligram of protein, acid phosphatase activity as micrograms of PO_4^{-3} produced per minute per milligram of protein, malate dehydrogenase activity as nanomoles of NADH oxidized per minute per milligram of protein, succinate dehydrogenase activity as nanomoles of succinate utilized per minute per milligram of protein, and adenylate kinase activity as nanomoles of NADP reduced per minute per milligram of protein [7]. Leakage of mitochondrial enzymes was measured in samples exposed to endotoxin for 45 min at 37°C [10]. Oxygen uptake by mitochondria was measured at 30°C with a Gilson oxygraph equipped with a

Clark electrode. Respiratory control ratios were expressed as nanogram atoms of oxygen consumed per minute per milligram of mitochondrial protein in the presence of ADP to the nanogram atoms of oxygen consumed per minute per milligram of mitochondrial protein in the absence of ADP. Coupling was measured as the ratio of micromoles of ADP utilized to microatoms of O_2 consumed [10]. The endotoxin used throughout was *Escherichia coli* 0127:B8 lipopolysaccharide, isolated by the Westphal procedure (Difco Laboratories, Detroit, MI).

RESULTS

Action of Endotoxin on Lysosomal and Mitochondrial Enzyme Activities

The activities of lysosomal β-glucuronidase or acid phosphatase become elevated in cultured cells exposed to 10 μg of endotoxin/ml for 4 hr and those of mitochondrial malate dehydrogenase or succinate dehydrogenase decrease within 2 hr. It has not been established whether it is necessary for the cells to be exposed to endotoxin continuously during this period or whether the cells, once exposed to endotoxin, are irreversibly committed to effect these metabolic changes. The lysosomal enzymes in primary cultures of perfused C3H/HeDub mouse liver cells did not increase unless endotoxin was present for several hours, indicating that endotoxin probably did not bind irreversibly to the cells and did not trigger irreversible metabolic alterations affecting lysosomal enzyme activity. Mitochondrial enzyme activities were decreased in cells exposed to endotoxin for only 1 hr and then incubated for 3 more hr without endotoxin, indicating that either endotoxin bound irreversibly to a constituent of the cell or triggered events affecting mitochondrial enzyme activities (Table I). Although the action of endotoxin on the vulnerable target may have occurred during the 1-hr exposure period, another hour is required for the perturbation to become manifested. These results also establish that a rise in lysosomal enzyme activities is not an inevitable consequence of impaired mitochondrial functions.

Responses of Beige Mice

The granulocytes of the Beige mouse have a defect in chemotaxis and several cell types, including leukocytes, contain giant lysosomes [12]. Beige mice and corresponding C57B1/6J mice are equally susceptible to endotoxin; a dose of 20 mg/kg killed about half of the treated animals. The lysosomal β-glucuronidase activity of primary cultures of liver cells from Beige mice was not elevated within 4 hr, even by 100 μg of endotoxin/ml. The lysosomal enzyme activity of liver cells prepared from other mouse strains doubled within 4 hr when exposed to 10 μg of endotoxin/ml. The mitochondrial malate dehydrogenase activity of endotoxin-treated liver cells from Beige mice was lower than that for untreated

TABLE I. Dependence of Enzyme Activities on Time of Exposure of Cell Cultures of Perfused C3H/HeDub Mouse Liver to Endotoxin*

Hours incubated		Enzyme activity				
With endotoxin[a]	Without endotoxin	Hexokinase	β-Glucuronidase	Acid phosphatase	Malate dehydrogenase	Succinate dehydrogenase
0	4	72 ± 12	3.3 ± 0.6	186 ± 24	321 ± 43	281 ± 17
1	0	78 ± 13	3.3 ± 0.4	193 ± 15	295 ± 24	304 ± 22
1	3	71 ± 12	3.1 ± 0.3	179 ± 12	196 ± 23[b]	191 ± 4[b]
2	0	67 ± 15	3.3 ± 0.3	164 ± 11	193 ± 25[b]	222 ± 16
2	2	64 ± 15	2.8 ± 0.3	182 ± 18	148 ± 37[b]	121 ± 13[b]
3	1	76 ± 19	3.2 ± 0.4	305 ± 28[b]	144 ± 21[b]	155 ± 12[b]
4	0	71 ± 4	5.0 ± 0.3[b]	327 ± 41[b]	180 ± 44[b]	194 ± 15[b]

*The medium was replaced with new medium without endotoxin after the period indicated. The cells were harvested after 4 hr, disrupted, and fractionated. Hexokinase was assayed in the 20,000g supernatant fraction, β-glucuronidase and acid phosphatase in the 20,000g sediment, and malate dehydrogenase and succinate dehydrogenase in the 10,000g sediment.

[a]*E. coli* 0127:B8 (Westphal) lipopolysaccharide 10 μg/ml.

[b]Significantly different from the untreated control at $P < 0.05$. Means ± standard deviations.

TABLE II. Effects of Endotoxin on Lysosomal β-Glucuronidase and Mitochondrial Malate Dehydrogenase Activities of Primary Cell Cultures of Perfused Liver from Beige Mice*

Endotoxin[a] (μg/ml)	Enzyme activities	
	β-Glucuronidase	Malate dehydrogenase
None	1.93 ± 0.16	261 ± 21
10	1.99 ± 0.15	243 ± 19
50	1.96 ± 0.16	186 ± 14[b]
100	1.93 ± 0.18	144 ± 11[b]

*Liver cells from Beige mice were incubated for 4 hr at 37°C in modified Eagle's medium, without added serum, and with the indicated amount of endotoxin.
[a]*E. coli* 0127:B8 (Westphal) lipopolysaccharide.
[b]Significantly different from the untreated control at P < 0.05. Means ± standard deviations.

TABLE III. Endotoxin-enhanced Leakage of Enzyme Activities from Mitochondrial Fractions of Beige Mice*

Endotoxin (μg/ml)	Enzyme activities[a]					
	Malate dehydrogenase		Succinate dehydrogenase		Isocitrate dehydrogenase	
	(total)	(% leaked)	(total)	(% leaked)	(total)	(% leaked)
None	192	7.2	137	8.1	221	9.1
5	205	11.7[b]	124	11.3	200	12.6
10	214	16.0[b]	134	15.8[b]	221	13.8[b]
25	212	26.2[b]	166	13.0[b]	221	13.9[b]

*Isolated mitochondria were incubated with *E. coli* 0127:B8 (Westphal) lipopolysaccharide for 45 min at 37°C.
[a]The specific activity of the total preparation (granular + soluble fractions) and the percentage of the activity leaked into the soluble fraction are reported.
[b]Significantly different from the control without endotoxin at P < 0.05.

liver cells, but 50 μg of endotoxin/ml was required (Table II). These results indicate that liver cells of Beige mice bind less endotoxin, or allow less endotoxin to reach the vulnerable target, or that the target is less responsive to endotoxin. In order to assess the latter proposal, isolated mitochondria from Beige mice were treated with endotoxin, and the loss of marker enzymes from the granular fraction was measured. The amount of malate dehydrogenase activity that leaked into the soluble fraction increased progressively with the endotoxin concentration. Loss of succinate dehydrogenase and isocitrate dehydrogenase activities was enhanced by endotoxin (Table III). Accordingly, endotoxin can interact directly on the isolated mitochondrial fraction leading to changes in the integrity of the mitochondria. These results indicate that the susceptibility of mitochondria to

TABLE IV. Endotoxin-induced Loss of Respiratory Control (RCR) in Mitochondria from Beige Mouse Liver Cells

Endotoxin (μg/ml)[a]	State 3[b]	State 4[b]	RCR	ADP/O
None	59.9	32.1	1.9	2.6
5	58.9	32.7	1.8	2.1
10	48.0	34.2	1.4[c]	2.0
25	32.1[c]	26.3[c]	1.2[c]	2.3
50	22.0[c]	22.0[c]	1.0[c]	—

[a]*E. coli* 0127:B8 (Westphal) lipopolysaccharide was added 5 min before the substrate (glutamate).
[b]Nanograms of oxygen consumed per minute per milligram of protein with substrate and ADP (state 3) or substrate only (state 4).
[c]Statistically different from the untreated control at $P < 0.05$.

the action of endotoxin is not altered in Beige mice, but that the binding, uptake, or translocation of endotoxin is altered.

The capability of isolated mitochondria from Beige mice to carry out state 4 respiration (with glutamate, without ADP) was not markedly altered by endotoxin (Table IV). These data agree with earlier results and affirm that endotoxin is not acting like an uncoupling agent. State 3 respiration (with glutamate and ADP) is impaired by endotoxin, thereby confirming that the mitochondria of Beige mice are susceptible to endotoxin. Respiratory control is reduced by endotoxin, indicating that the integrity of the mitochondria is perturbed; however, oxidative phosphorylation (ADP/O) is unchanged.

DISCUSSION

Relating Effects on Cells with Effects on Animals

The concentration of endotoxin that elicits a marked decrease in mitochondrial enzyme activities in cultured cells (10 μg/ml) is comparable to the endotoxin dose that kills 50% of treated mice (10–20 μg of endotoxin/g of mouse weight). Furthermore, endotoxin causes a substantial change in mitochondrial enzyme activities in cultured cells within 2–4 hr, which corresponds to the time that animals display loss of thermal regulation and pactamycin-treated mice die after administration of endotoxin intravenously [2]. It is now well established that endotoxin can affect cellular metabolic processes directly and that all of the cellular effects of endotoxin cannot be attributed to immunologic reactions, changes in hormone levels, or humoral mediators.

Gram-negative bacterial endotoxin impairs mitochondrial function when injected into animals, added to cells in culture, or added to subcellular fractions [1]. Although endotoxin affects the activity of isolated mitochondria, it is not definitively established that the interaction of endotoxin with the mitochondria per se is the causal early event in endotoxicosis. It is conceivable that endotoxin acts first on the plasma membrane, thereby triggering several diverse series of reactions. In order to identify the site of some of the early determinative alterations elicited by endotoxin, the effects of endotoxin on liver cells in primary cultures and subcellular fractions of liver from endotoxin-susceptible mice, from mice rendered resistant to endotoxin by prior treatment with endotoxin, and from genetically endotoxin-refractory mice have been examined.

Liver cells from normal endotoxin-susceptible mice have reduced activities of mitochondrial enzymes after being exposed to endotoxin for a few hours. Liver cells from endotoxin-refractory mice, whether rendered resistant by prior treatment with endotoxin or from the genetically resistant C3H/HeJ mouse, exhibit normal mitochondrial enzyme activity after being exposed to endotoxin. Isolated mitochondria from endotoxin-susceptible and endotoxin-refractory mice are sensitive to endotoxin, indicating that resistance is a cellular and not an organellar attribute [13]. Several investigators have reported that cells of C3H/HeJ and C3H/HeN mice bind endotoxin equally well [14]. These observations indicate that, although binding of endotoxin should be a prerequisite for a response, binding is not by itself sufficient to cause a response. The mechanism by which endotoxin is bound to the cell membranes is largely unknown. Any membrane constituent with a high affinity for the lipid moiety of endotoxin may bind it. Binding to such high-affinity, but nonspecific, sites on membranes may be sufficiently strong to obscure specific binding to a specific endotoxin receptor. This phenomenon may account for the failure of most, but not all, workers to detect differences in binding of endotoxin by cells from endotoxin-susceptible and -resistant mice [15]. The possibility that resistance in the C3H/HeJ mouse reflects altered internalization of endotoxin or impaired triggering of crucial responses on the plasma membrane must also be considered. Alternatively, resistance in the C3H/HeJ mouse may involve impaired translocation of the toxophore or effector through the cytoplasm to the ultimate target.

Relating Effects on Liver with Molecular Events

Numerous metabolic activities in liver are altered after endotoxicosis. Such changes include depletion of carbohydrate reserves, inhibition of particular enzyme induction, and inhibition of gluconeogenesis [16]. Endotoxin induces a transient hyperglycemia followed by a marked hypoglycemia. The early hyperglycemia is the result of increased hepatic glucose output due to enhanced glycogenolysis. The hypoglycemia develops as a consequence of both elevated uptake of glucose by peripheral tissue and the failure of the liver to compensate for this by augmented glucose output [17]. Liver damage initiated by endotoxin leads to markedly elevated levels of transaminase activities in serum within a few hours [3] and to decreased levels of microsomal mixed function oxygenase activity [5].

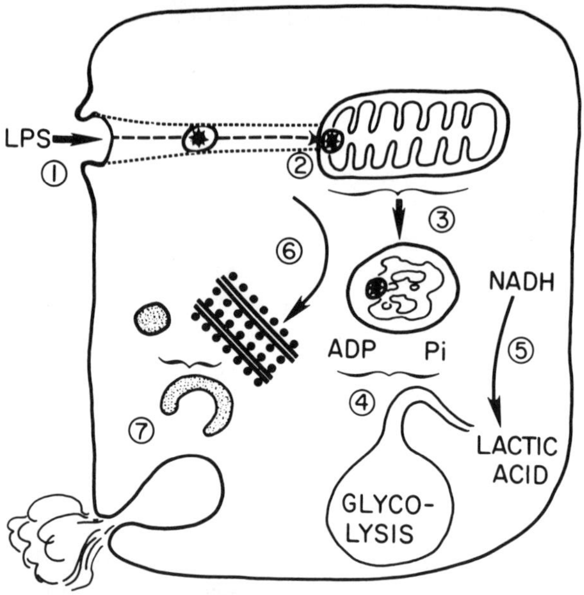

Fig. 1. Cellular basis for the action of endotoxin. (1) Endocytosis and (2) translocation of endotoxin to the mitochondrion. (3) The proton gradient is destroyed, causing (4) ADP and NADH to accumulate the cytosol, and (5) enhanced glycolysis. (6) Lysosomal enzymes are induced, autophagy is stimulated and (7) lysosomal hydrolases are released.

Livers of normal mice trap in 1 hr more than 80% of endotoxin injected intravenously [18]. Willerson et al [19] have concluded that endotoxin accumulates in the mitochondrial fraction of liver and spleen of rats and mice. Most of the endotoxin is associated with parenchymal cells [20]. A central question in addressing the role of the liver in the pathophysiology of endotoxin shock is whether parenchymal cells are affected directly by endotoxin or whether metabolic changes observed in endotoxemia are mediated through soluble products released from macrophages after phagocytosis of endotoxin [16].

Elucidation of the molecular mechanism of action of a toxin can be achieved only by characterizing the active site of the toxin molecule, by defining the receptor, by determining the nature of the initial interaction between the toxin and its receptor, and by relating the initial events to the sequelae of cellular alterations that culminate in characteristic symptoms in susceptible animals. Although the molecular bases of the toxicity of gram-negative bacterial endotoxin is not resolved definitively, a substantial body of information now exists so that reasonable models can be formulated.

Results to date indicate that lipid A represents the endotoxically active center of the molecule, whereas the polysaccharide portion serves as a solubilizing carrier for an otherwise insoluble lipid [1]. Complexes of myristic acid with bovine serum albumin, myristic acid with concanavalin A, dimethyl myristamide with concanavalin A, and sphingosine with bovine serum albumin possess endotoxic activity [1, 21]. Springer et al [14] have isolated glycerophosphatides from platelets and leukocytes that combine with endotoxin. These observations are consistent with the proposition that the lipid A moiety binds to a lipid receptor or inserts into lipoidal constituent of the cell [1]. Cardiolipin is a distinctive constituent of the mitochondrial membrane and an intact mitochondrial membrane is necessary for oxidative phosphorylation. It is conceivable that cardiolipin is a component of the endotoxin receptor or that cardiolipin is an analog of lipid A.

The following model for the cellular and molecular bases for endotoxin's action is proposed: (a) Endotoxin is taken into the cell by endocytosis, (b) endotoxin is translocated through the cytoplasm in the endocytic vacuole, (c) the endocytic vacuole may fuse with a primary lysosome, (d) the endotoxic toxophore is transferred to its specific receptor in the mitochondrial membrane, (e) the mitochondrial proton gradient is destroyed, and (f) ADP and NADH accumulate in the cytosol, leading to enhanced glycolytic metabolism (Fig. 1). These cellular alterations culminate in the characteristic symptoms of endotoxicosis: hypoglycemia, disseminated intravascular coagulation, and severe circulatory hypotension. These latter changes reflect the activities of lysosomal enzymes, in particular, protease activity and humoral mediators released by injured host cells.

REFERENCES

1. Bradley SG: Cellular and molecular mechanisms of action of bacterial endotoxins. Ann Rev Microbiol 3:67, 1979.
2. Bradley SG, Bond JS: Toxicity, clearance and metabolic effects of pactamycin in combination with bacterial lipopolysaccharide. Toxicol Appl Pharmacol 31:208, 1975.
3. Munson AE, Drummond DC, Adams AC, Bradley SG: Enhanced toxicity for mice of combinations of bacterial lipopolysaccharide and vincristine. Antimicrob Agt Chemother 9:840, 1976.
4. Marecki NM, Bradley, SG, Munson AE, Drummond DC: Effect of bacterial lipopolysaccharide, lipid A, and concanavalin A on lethality of 5-fluorouracil for mice. Toxicol Appl Pharmacol 31:83, 1975.
5. Rose WC, Bradley SG, Lee IP: Enhanced toxicity for mice of vincristine and other chemotherapeutic agents with *Salmonella typhosa* endotoxin and *Pseudomonas aeruginosa*. Antimicrob Agt Chemother 1:489, 1972.
6. McGivney A, Bradley SG: Effects of bacterial endotoxin on lysosomal enzyme activities of human cells in continuous culture. J Reticuloendothel Soc 23:223, 1978.
7. McGivney A, Bradley SG: Effects of bacterial endotoxin on lysosomal and mitochondrial enzyme activities of established cell cultures. J Reticuloendothel Soc 26:307, 1979.
8. McGivney A, Bradley SG: Enhanced lysosomal enzyme activity in mouse cells treated with bacterial endotoxin in vitro. Proc Soc Exp Biol Med 155:390, 1977.

9. Mørland B: Studies on selective induction of lysosomal enzyme activities in mouse peritoneal macrophages. J Reticuloendothel Soc 26:749, 1979.
10. McGivney A, Bradley SG: Action of bacterial endotoxin and lipid A on mitochondrial enzyme activities of cells in culture and subcellular fractions. Infect Immun 25:664, 1979.
11. McGivney A, Bradley SG: Action of bacterial lipopolysaccharide on the respiration of mouse liver mitochondria. Infect Immun 27:102, 1980.
12. Gallin JI, Bujak JS, Patten E, Wolff SM: Granulocyte function in the Chediak–Higashi syndrome of mice. Blood 43:201, 1974.
13. McGivney A, Bradley SG: Susceptibility of mitochondria from endotoxin-resistant mice to lipopolysaccharide. Proc Soc Exp Biol Med 163:56, 1980.
14. Springer GF, Adye JC, Mergenhagen SE, Rosenstreich DL: Endotoxin interactions with phosphatides from human leukocytes and platelets, and with lymphoid cells of susceptible and resistant mice. In Schlessinger D (ed): "Microbiology 1977." Washington, D.C.: Am Soc Microbiol, 1977, p 326.
15. Nygren H, Dahlén G, Möller G: Bacterial lipopolysaccharides bind selectively to lymphocytes from lipopolysaccharide high-responder mouse strains. Scand J Immunol 10:555, 1979.
16. Berry LJ: Bacterial toxins. CRC Crit Rev Toxicol 5:239, 1977.
17. Wolfe RR, Elahi D, Spitzer J: Glucose and lactate kinetics after endotoxin administration in dogs. Am J Physiol 232:E180, 1977.
18. Walker RI, Snyder SL: Endotoxemia in irradiated mice. In Rosenberg P (ed): "Toxins: Animal, Plant and Microbial." New York: Pergamon, 1978, p 933.
19. Willerson JT, Trelstad RL, Pincus T, Levy SB, Wolff SM: Subcellular localization of *Salmonella enteriditis* endotoxin in liver and spleen of mice and rats. Infect Immun 1:440, 1970.
20. Zlydaszyk JC, Moon RJ: Fate of ^{51}Cr-labelled lipopolysaccharide in tissue culture cells and livers of normal mice. Infect Immun 14:100, 1976.
21. Bradley SG: Endotoxic activity of complexes of myristic acid and proteins. Proc Soc Exp Biol Med 151:267, 1976.

Direct and Indirect Effects of Endotoxin on Mitochondrial Function

Leena Mela

A multitude of cellular alterations indicative of cell damage and vital organ dysfunction have been identified in sepsis and endotoxemia. The role of direct effects of endotoxin in the precipitation of cellular damage, however, is not clear. We have used mitochondrial function as a sensitive indicator of irreversible intracellular damage and dysfunction. Mitochondrial function was analyzed after in vivo *Escherichia coli* endotoxemia and live *E. coli* septicemia. Comparisons of cellular effects of endotoxin in vivo with direct endotoxin effects on isolated mitochondria in vitro provide information about direct endotoxin effects in the precipitation of cell damage and, eventually, death.

METHODS

In vivo endotoxemia was induced in guinea pigs by an intraperitoneal injection of a dose of *E. coli* endotoxin, which caused 60% mortality at 24 hr and 100% mortality by 3 days [1]. This preparation was obtained from Difco Laboratories (Detroit, MI) and was the lipopolysaccharide B (Boivin extraction), strain 0111:B4. In treated animals, 100 mg/kg of methylprednisolone sodium succinate (MPSS) was given simultaneously with endotoxin.

Live *E. coli* septicemia was induced in dogs to simulate in vivo sepsis. An LD_{100} dose (at 24 hr) of live *E. coli* organisms (about 10^{10}/kg) was given intravenously [2]. Treated animals received a total of 60 mg/kg of MPSS as infusion starting at 15 min after the injection of bacteria. Gentamicin was given 60 min after bacterial injection, followed by two additional injections during the first 6 hr. A total of 6.5 mg/kg of gentamicin was used.

In vitro endotoxin effects on mitochondrial function were determined in suspensions of freshly isolated normal kidney and brain mitochondria. For these studies chromatographically purified preparations of *E. coli* endotoxin were used (E. coli 0111:B4, Westphal extraction, from Calbiochem, San Diego, CA). When used, MPSS was added to the mitochondrial suspensions prior to the addition of endotoxin at concentrations ranging from 5 to 40 μg/mg of protein.

Pathophysiological Effects of Endotoxins at the Cellular Level, pages 15–21

TABLE I. Effect of Terminal In Vivo Endotoxemia on Rat Liver Mitochondrial Function of *E. coli* Endotoxin, LD_{70} at 24 HR

Measured parameter	Unit	Control	Endotoxic
State 4 respiration			
Glutamate and malate	nmole O_2/min/mg	5.0	10.0
State 3 respiration			
Glutamate and malate and ADP	nmole O_2/min/mg	35	15
Respiratory control			
State 3/state 4		7.0	1.5
ADP utilization rate	nmole/min/mg	160	0
ATP hydrolysis rate	nmole/min/mg	63	30
Ca^{2+} uptake rate	nmole/min/mg	400	150–0
Mg^{2+} concentration	nmole/mg	30	18

Kidney and brain mitochondria were isolated from guinea pigs from control and from endotoxic and septic animals at various times after the injection of endotoxin or bacteria [1–3]. All mitochondrial samples were analyzed for respiratory activities in state 4 (resting state) and state 3 (active state, in the presence of ADP), respiratory control, and calcium accumulation rates. In addition, absorption spectra of the respiratory chain components were determined. Concentrations of cytochromes $a + a_3$, c, and b were calculated. Mitochondrial respiratory and calcium transport activities were standardized per milligram of mitochondrial total protein and per mole of cytochrome oxidase.

RESULTS AND DISCUSSION

As previously reported by us, *E. coli* endotoxemia in animals receiving no treatment induces inhibition of mitochondrial function in a time- and dose-dependent manner [1,2]. Table I summarizes liver mitochondrial data from 50 animals (rats) 18–24 hr after an injection of *E. coli* endotoxin. It should be noted that the mitochondrial parameters specifically affected by in vivo endotoxemia have been determined to be the energy-linked reactions involved in ATP synthesis (adenine nucleotide translocase and the ATPase–ATP synthetase) and calcium ion accumulation [1,4]. The changes involving ATP synthesis reactions are indicated by the decline of respiratory rates in state 3 (+ADP). This is also reflected in the decline of respiratory control ratios, in the lack of ADP-induced redox response of the cytochromes, and in the lowered rates of uncoupler-stimulated ATPase activity. The inhibition of mitochondrial capacity to accumulate calcium is indicated by the lowered initial rates of calcium transport as well as by the decreased amount of total calcium accumulated per milligram of mito-

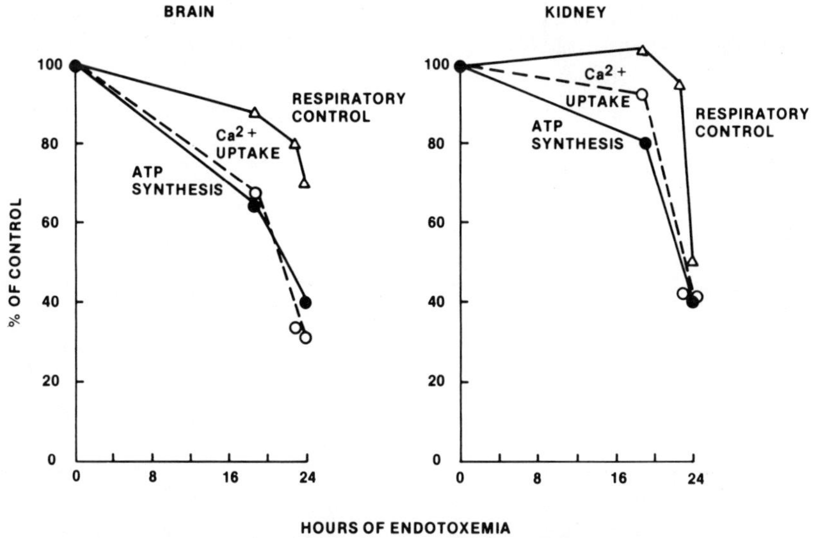

Fig. 1. Mitochondrial function after in vivo *E. coli* endotexemia. An LD_{70} dose of endotoxin (24 hr. mortality) was given intraperitoneally at time 0. At 18 or 24 hours brain and kidney mitochondria were isolated and their function analyzed in vitro. ATP synthesis rates in nmoles/min/mg protein (●), respiratory controls (△) and Ca^{2+} transport rates (○) were analyzed and calculated as % of control, as described previously (1).

TABLE II. Brain and Kidney Mitochondrial Respiratory Activities after 8–24 Hr of Live *E. coli* Septicemia in the Dog*

State 3 respiration[a]	Control	Septicemic
Brain	440	235
Kidney	410	400

*Data are from Mela [2]. This work was performed in collaboration with Dr. Lerner Hinshaw.
[a]Moles of O_2 per minute per mole of cytochrome oxidase.

chondrial protein. Figure 1 summarizes similar data from mitochondria isolated after 24 hr of endotoxemia from two vital organs, the brain and the kidney. These data illustrate the different sensitivities of the vital organs to in vivo endotoxemia indicating early inhibition of brain mitochondrial function followed later by the liver and the kidney. Similar time differences in the deterioration of mitochondrial function were found in live *E. coli* septicemia in the dog. In those experiments, brain mitochondrial function deteriorated prior to any decline in kidney mitochondrial function [2]. These data are summarized in Table II.

We have utilized the studies of protective effects of glucocorticoids in endotoxemia and glucocorticoid–antibiotic combination treatment in septicemia

TABLE III. Protection by Early Glucocorticoid Treatment in Endotoxemia and Septicemia in Experimental Animals*

Treatment	Mortality (%)	Brain mitochondrial ATP synthesis rate (% of control)	Total amount of drug
E. coli endotoxemia of the guinea pig			
Untreated	60	45	0
Treated	0	108	100 mg/kg of methylprednisolone Na-succinate
Live *E. coli* septicemia of the dog			
Untreated	100	55	0
Treated	0	106	60 mg/kg of methylprednisolone Na-succinate and 6.5 mg/kg of gentamicin

*Data are from Mela [2]. Live *E. coli* septicemia experiments were preformed collaboratively with Dr. Lerner Hinshaw.

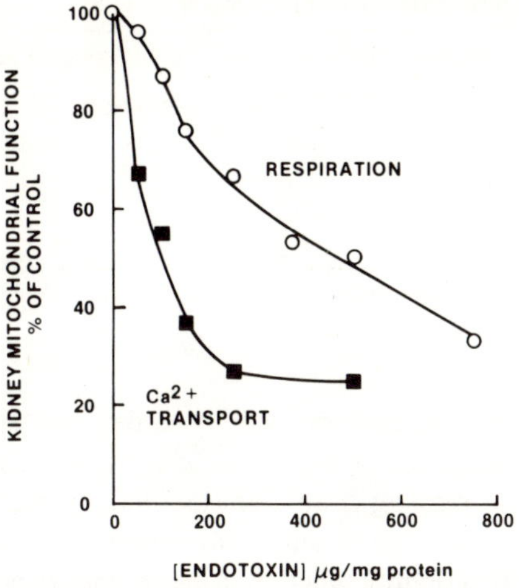

Fig. 2. In vitro effects of *E. coli* endotoxin on mitrochondrial activity. Normal isolated kidney mitochondrial respiratory activity in state 3 (in the presence of substrate and ADP) (○) and Ca^{2+} transport activity (■) were determined in the presence of varying concentrations of endotoxin and calculated as percentages of control (no endotoxin), as previously described (1).

against the lethality in circulatory shock as a means to determine the role of mitochondrial functional deterioration in vital organ failure in sepsis [2]. As Table III illustrates, early treatment in both cases prevents mortality. It has also been shown previously that early glucocorticoid treatment protects against lowered vital organ blood flow in endotoxemia and prevents the microcirculatory alterations induced by granulocyte and platelet aggregation [5–7]. The improved perfusion of vital organs, such as the brain and the kidney, might contribute to the protection against metabolic failure at the mitochondrial level seen in treated animals.

A pattern of mitochondrial inhibition similiar to that found after endotoxemia has been shown to occur after complete and partial tissue ischemia [3]. This finding indirectly supports the suggestion that endotoxin-induced mitochondrial deterioration is partially due to ischemia of the organ. The data on mitochondrial function after in vivo endotoxemia alone, however, cannot eliminate the possibility that direct effects of endotoxin on mitochondrial membranes and function might play a role in the irreversible functional failure of cellular energy metabolism in sepsis. Our studies of direct in vitro endotoxin effects on mitochondrial function were designed to clarify this point. Figure 2 illustrates a titration of mitochondrial respiratory activity in state 3 (+ADP) and calcium accumulation by mitochondria as a function of increasing concentrations of added endotoxin in the suspension. A 50% inhibition of kidney mitochondrial function is achieved at an endotoxin concentration ranging from 286 μg/mg of protein to over 500

TABLE IV. Endotoxin Accumulation in Vital Organs In Vivo

Organ	In vivo accumulation percentage of total amount injected (%)[a]	(amount in mg)[b]	Amount needed for 50% inhibition of mitochondrial function[b] (mg)
Liver	88.4	2.7	
Kidney	0.1	0.003	5–8
Brain	<2.8	0.009	>6.6

[a]Data are from Zlydaszyk and Moon [8].
[b]Data are from Mela, unpublished observations.

μg/mg of protein. Calcium transport activity is the most sensitive indicator of mitochondrial inhibition. Similar inhibition of brain mitochondrial function requires more endotoxin, ranging from 1000 to 2000 μg/mg of protein.

These findings are quite revealing: The amount of endotoxin required for direct inhibition of mitochondrial function in the absence of any other contributing factors is very high. Taking into consideration the total amount of mitochondrial protein collected during the isolation procedure from each organ, we calculated the amount of endotoxin to be accumulated in each organ for a 50% inhibition of mitochondrial function to occur. These results are reported in Table IV. This table also presents data by Zlydaszyk and Moon on the distribution of endotoxin injected in vivo into the vital organs of the mouse [8]. A comparison of these two sets of data indicates that a sole direct effect of phenol-extracted biologically unmodified endotoxin on mitochondrial function as a cause of irreversible metabolic failure cannot be considered. In addition, the fact that glucocorticoids provide no protection against direct endotoxic effects on mitochondrial function in vitro supports the same conclusion (data not shown). Thus, other factors need to be considered. Lowered tissue perfusion as well as formation of possible secondary toxins or mediators, activation of lysosomal enzymes, lowered tissue pH, and altered sodium, potassium, magnesium, and calcium concentrations need be taken into consideration. The findings described above also suggest that liver, brain, and kidney mitochondrial changes most likely are not the primary causes of cell death in sepsis and endotoxemia. Rather, they appear secondary, caused by several contributing cellular alterations. The mitochondrial changes, however, might be directly related to irreversibility in sepsis. Our data on glucocorticoid protection in endotoxemia demonstrate the close direct relationship between mitochondrial damage and high mortality. These data are shown in Figure 3. With glucocorticoids injected simultaneously with endotoxin, no mortality was found at 24 hr. Both brain and kidney mitochondria exhibited normal energy-linked functions. When glucocorticoids were delayed by 1 or 2 hr, mortality increased, as did the inhibition of mitochondrial function.

We, thus, feel that if inhibition of mitochondrial function can be prevented, vital organ failure and death in sepsis can be abolished. Although inhibition of

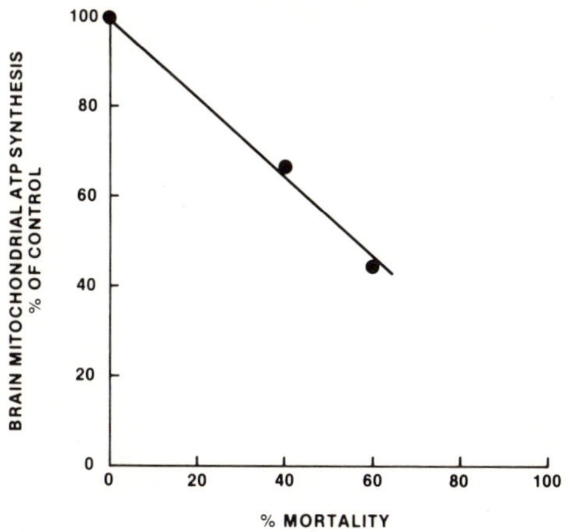

Fig. 3 The points indicate three groups of animals (n = 10–15 in each group). Zero mortality is achieved in animals treated with methylprednisolone sodium succinate simultaneously with endotoxin; 40% mortality if treatment is delayed 2 hr; 60% mortality is seen in untreated animals.

mitochrondrial function cannot be induced by biologically unmodified endotoxin in vitro, endotoxin in vivo can irreversibly inhibit mitochondrial function via mediators or following its biological modification. These mitochondrial changes are probably key events in the fatal cellular dysfunction in septicemia and endotoxemia.

REFERENCES

1. Mela L, Bacalzo LV, Miller LD: Defective oxidative metabolism of rat liver mitochondria in hemorrhagic and endotoxic shock. Am J Physiol 220:571, 1971.
2. Mela L: Reversibility of mitochondrial metabolic response to circulatory shock and tissue ischemia. Circ Shock 1:61, 1979.
3. Rehncrona S, Mela L, Siesjo BK: Recovery of brain mitochondrial function in the rat after complete and incomplete cerebral ischemia. Stroke 10:437, 1979.
4. Mela L: Mitochondrial metabolic alterations in experimental circulatory shock. In Urbaschek B, Urbaschek R, Neter E (eds): "Gram-negative Bacterial Infections and Mode of Endotoxin Actions." New York: Springer–Verlag, 1975, p 288.
5. Emerson TE, Bryan WJ: Regional cerebral blood flows in endotoxin shock with methylprednisolone treatment. Proc Exp Biol Med 156:378, 1977.
6. Balis JV, Paterson JF, Shelley SA, et al: Glucocorticoid and antibiotic effects on hepatic microcirculation and associated host responses in lethal gram-negative bacteremia. Lab Invest 40:55, 1979.
7. O'Flaherty JT, Craddock PR, Jacob H: Mechanism of anti-complementary activity of corticosteroids in vivo: Possible relevance in endotoxin shock. Proc Soc Exp Biol Med 154:206, 1977.
8. Zlydaszyk JC, Moon RJ: Fate of ^{51}Cr-labeled lipopolysaccharide in tissue culture cells and livers of normal mice. Infect Immunol 14:100, 1976.

Comparison of the In Vitro and In Vivo Effects of Gram-Negative Endotoxin on Cardiac Sarcoplasmic Reticulum and Myofibrillar Function

Stephen M. Krause and Michael L. Hess

The effects of gram-negative endotoxin on myocardial function can be divided into two subcategories: one dealing with a holistic effect on the heart and its subcellular components and the other dealing with the direct effects of endotoxin on the myocardium. In order to understand this problem better, this laboratory has directed its attention to the subcellular derangements which may occur as a result of in vivo endotoxemia or in vitro exposure to endotoxin. The two subcellular organelles utilized for this study are the sarcoplasmic reticulum (SR) and the contractile proteins. As demonstrated by Figure 1, the sarcoplasmic reticulum is both the source and sink for coupling calcium in the excitation–contraction coupling sequence [1] and is thus regulatory in nature. Upon stimulation of the myocardial cell, Ca^{2+} is released by the sarcoplasmic reticulum and results in activation of the myofibrils. Following contraction, Ca^{2+} is actively sequestered by the sarcoplasmic reticulum, and this results in relaxation. The contractile proteins are responsible for tension generation. With activation, calcium is bound to the regulatory protein, troponin; myosin ATPase is activated; ATP is hydrolyzed; and cross-bridge formation ensues. In the absence of calcium and the presence of ATP, cross-bridge formation is inhibited and relaxation results. Thus two major energy-dependent organelles within the cardiac cell that could be injured by either endotoxin alone or by endotoxin-shock processes are the sarcoplasmic reticulum and the myofibrils.

There is currently a large body of data incriminating myocardial dysfunction during the course of endotoxin shock. On gross examination, Kleinman et al [2] demonstrated significant subendocardial hemorrhagic necrosis which, when examined with the vital stain thioflavin-S, was found to be due to subendocardial under perfusion. Electron micrographs of the subendocardial surface demonstrate

Pathophysiological Effects of Endotoxins at the Cellular Level, pages 23 – 32

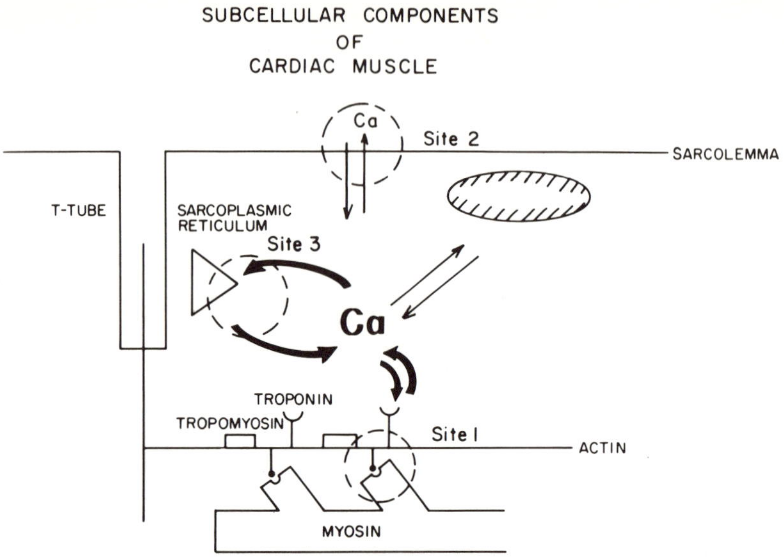

Fig. 1. A schematic representation of a myocardial cell depicting the role of calcium at the three major sites: site 1, the myofibril; site 2, the sarcolemma; and site 3, the sarcoplasmic reticulum.

a disruption of the contractile proteins and a depletion of glycogen, an injury process that spares the subepicardium. This would appear to incriminate an ischemic process as an etiology for this subendocardial necrosis which would, in turn, result in the documented myocardial failure observed during endotoxin shock.

The question that needs to be examined is whether endotoxin exerts this effect directly or indirectly. In order to try to answer this question, we studied the effects of endotoxin, both in vivo and in vitro, on the function of the sarcoplasmic reticulum and the myofibrils (contractile proteins).

METHODS

Endotoxin shock *(Escherichia coli,* 0.26:B6, Boivin extraction, Difco; 4mg/kg, N = 7) was induced in healthy 20- to 25-kg dogs and hemodynamically monitored for five hr following the previously described protocol utilized by this laboratory [3]. At the conclusion of the hemodynamic phase of the study, the hearts were removed, and the sarcoplasmic reticulum was isolated following the method of Hess [4] and the myofibrils isolated following the method of Krause et al [5]. The in vitro studies were performed on sarcoplasmic reticulum and myofibrils isolated from normal control hearts (N = 7) utilizing EGTA-dialyzed endotoxin in order to remove contaminating exchangeable calcium [6]. Statistical analysis was performed utilizing Student's t test with values of $P < 0.05$, indicating a significant difference.

RESULTS

In Vitro Studies

Figure 2 demonstrates the concentration dependency of endotoxin inhibition of sarcoplasmic reticulum calcium-uptake rates. Increasing the concentration of endotoxin from 0.024 to 0.24 mg/ml caused a significant linear depression of calcium-uptake velocities by 57%. As demonstrated by Figure 3, this effect can be potentiated by preincubating the sarcoplasmic reticulum with endotoxin (0.24 mg/ml). Preincubation of the sarcoplasmic reticulum with endotoxin at 37°C resulted in a linear increase in the depression of calcium uptake with a 72% loss in calcium-transport function with 30 min of preincubation. Figure 4 demonstrates the temperature dependence of endotoxin inhibition of calcium uptake by isolated cardiac sarcoplasmic reticulum. This figure demonstrates that the endo-

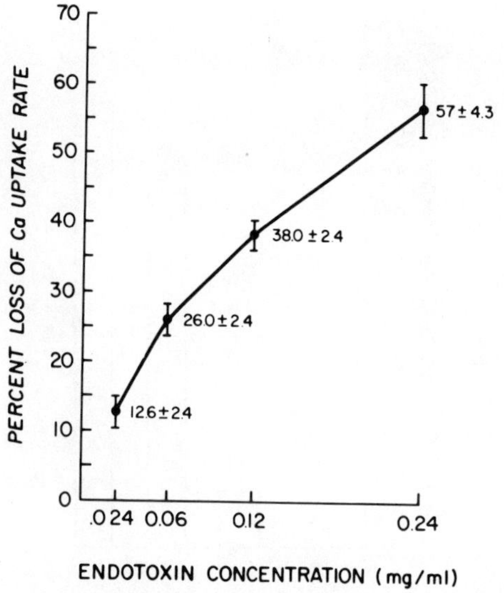

Fig. 2. The concentration dependency of endotoxin on calcium uptake of the sarcoplasmic reticulum (SR) in vitro expressed as a loss of ability to actively sequester calcium. Each point represents the mean and each bar represents one standard error of the mean.

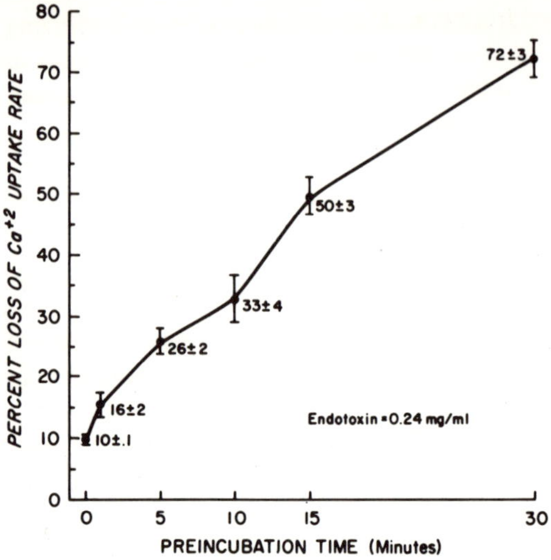

Fig. 3. The effect of preincubation time on calcium uptake rate of SR in vitro. All experiments performed at 37°C with an endotoxin concentration of 0.24 mg/ml. Each point represents the mean and each bar represents one standard error of the mean (reproduced by permission of Biochem Biophys Res Commun).

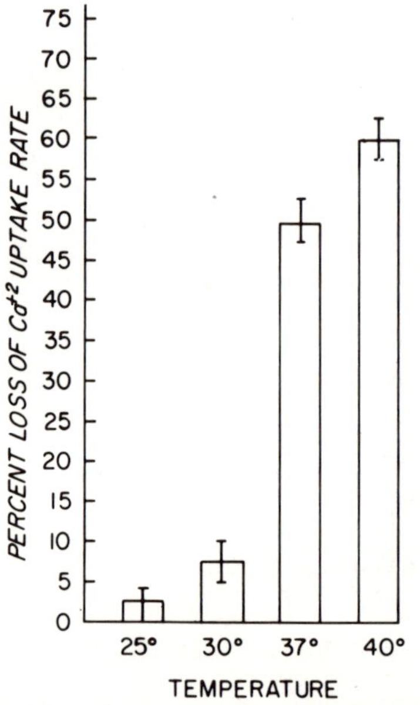

Fig. 4. Temperature dependency of calcium uptake rate in SR in vitro. All experiments performed with an endotoxin concentration of 0.24 mg/ml. Each bar represents the mean and each line represents one standard error of the mean (reproduced by permission of Myocard Biol).

toxin effect exhibits a marked temperature dependence in this in vitro system. At 25°C there is little if any inhibition noted; at 30°C this inhibition increased to 8%, but between 30 and 40°C there is a marked increase in the potency of endotoxin to inhibit sarcoplasmic reticulum calcium-uptake rates.

Having established that endotoxin inhibition of sarcoplasmic reticulum calcium transport is concentration, time, and temperature dependent, we next turned our attention to the component parts of the sarcoplasmic reticulum calcium-transport system. Figure 5 demonstrates that endotoxin is probably acting on the phospholipid portion of the sarcoplasmic reticulum. This figure demonstrates that the polyhydroxyl alcohols, sucrose and glycerol, are capable of significantly antagonizing the depressant effects of endotoxin. This protection is probably due to a coating of the phospholipids by the sugars, thus allowing them to remain intact. A further strengthening of this hypothesis comes from the measured Ca^{2+}–ATPase activity associated with calcium transport. Control Ca^{2+}–ATPase activity averaged 2.4 ± 0.4 μmole of P_i/mg-min, while endotoxin-treated sarcoplasmic reticulum averaged 2.8 ± 0.4 μmole of P_i/mg-min; $P > 0.2$. Thus, it appears that in vitro endotoxin uncouples calcium transport from ATP hydrolysis in isolated cardiac sarcoplasmic reticulum.

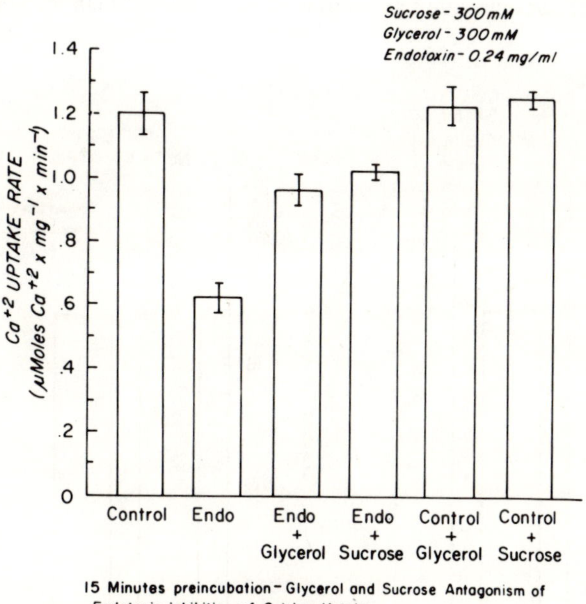

Fig. 5. The effect of 300 mM sucrose and 300 mM glycerol on SR calcium uptake in vitro with 0.24 mg/ml of endotoxin. Each bar represents the mean and each line represents one standard error of the mean (reproduced by permission of Myocard Biol).

The two other major enzyme systems that are involved in the regulation of myocardial contractility are the sarcolemmal Na^+–K^+ ATPase system and the myofibrillar Ca^{2+}–ATPase system. Figure 6 demonstrates that in vitro endotoxin has no effect on either the sodium–potassium ATPase system or the myofibrillar Ca^{2+}–ATPase system, again demonstrating that enzyme systems are relatively resistant to the direct effects of endotoxin, while phospholipid-mediated transport systems are quite sensitive to the effects of endotoxin. Based on the above data, it appears that if endotoxin can enter the myocardial cell, it is capable of exerting a direct negative inotropic effect by uncoupling calcium transport from ATP hydrolysis at the level of site 3, the sarcoplasmic reticulum. However, the significant unanswered question remains: Does endotoxin permeate the myocardial cell?

In Vivo Studies

An answer to the question posed above comes from the study of the effects of endotoxemia on myocardial subcellular function. Figure 7 demonstrates the effect of 5 hr of endotoxemia on the calcium-uptake velocity of isolated sarcoplasmic reticulum. The depression of calcium uptake following in vivo exposure of the myocardium to endotoxin is similar to the in vitro studies, but the subendocardial sarcoplasmic reticulum is more depressed than the subepicardial

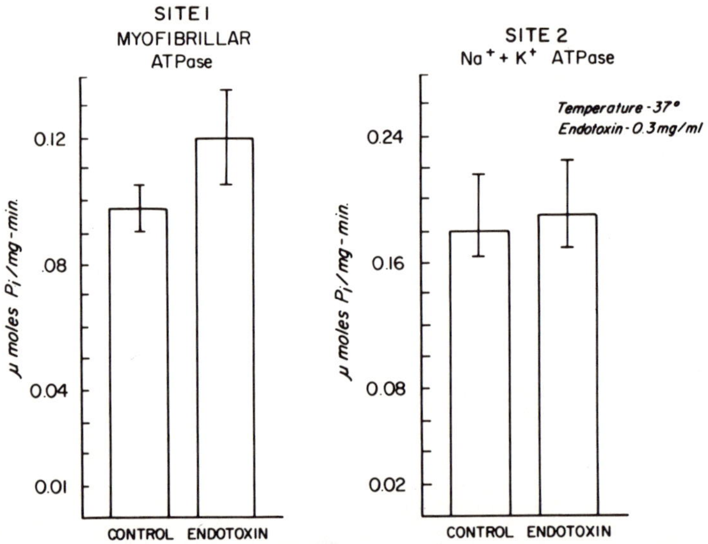

Fig. 6. The in vitro effect of endotoxin on enzyme activity at the myofibrillar and sarcolemmal sites. Each bar represents the mean and each line represents one standard error of the mean.

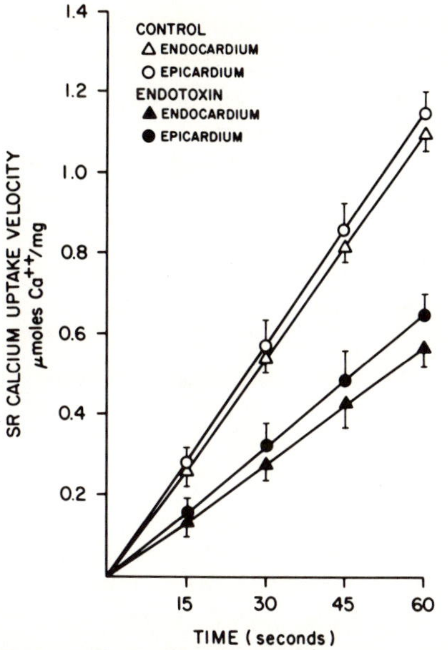

Fig. 7. The effect of endotoxin shock on SR calcium uptake velocity expressed as μmoles of Ca^{2+} per milligram. From the endocardial and epicardial surface of the myocardium, each point represents the mean and each line represents one standard error of the mean.

sarcoplasmic reticulum. Another striking difference is observed when one examines the Ca^{2+}-stimulated, Mg^{2+}-dependent ATPase activity of the isolated sarcoplasmic reticulum (Fig. 8). In marked contrast to the in vitro studies, the Ca^{2+}–ATPase activity of the isolated sarcoplasmic reticulum from the in vivo studies demonstrates a 50% depression of activity from 1.2 μmole of P_i/mg-min to 0.6 μmole of P_i/mg-min ($P < 0.05$), incriminating both a depression and uncoupling of the calcium-transport system.

Figure 9 presents the evaluation of the functional integrity of the contractile proteins as evaluated by pCa–myofibrillar ATPase curves. In marked contrast to the in vitro studies, there is a depression of myofibrillar ATPase activity, once again demonstrating an unequal distribution with the endocardial surface demonstrating the greatest depression. This depression is not only due to a decrease in calcium availability, but because the curves are generated over a range of free calcium concentrations known to exist in the cell [7], this would incriminate a depression in contractile protein function. This is best seen at the endocardial surface where maximal ATPase activity is significantly depressed from a control value of 0.14 to a shocked value of 0.11 μmole of P_i/mg-min. The curve derived from the epicardial myofibrillar ATPase activity is slightly different. There is

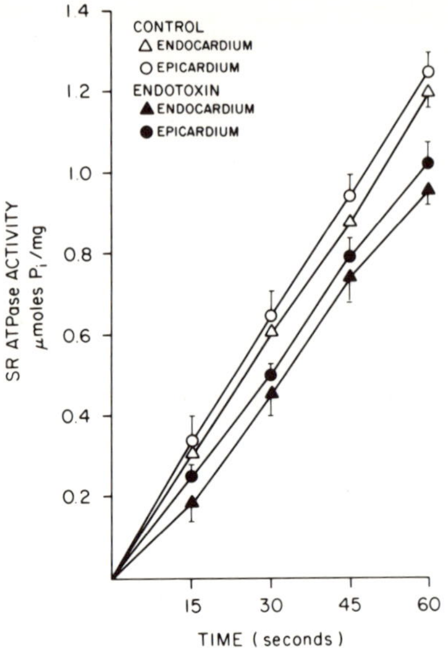

Fig. 8. The effect of endotoxin shock on SR ATPase activity expressed as μmoles of P_i per milligram. From the endocardial and epicardial surface of the myocardium, each point represents the mean and each line represents one standard error of the mean.

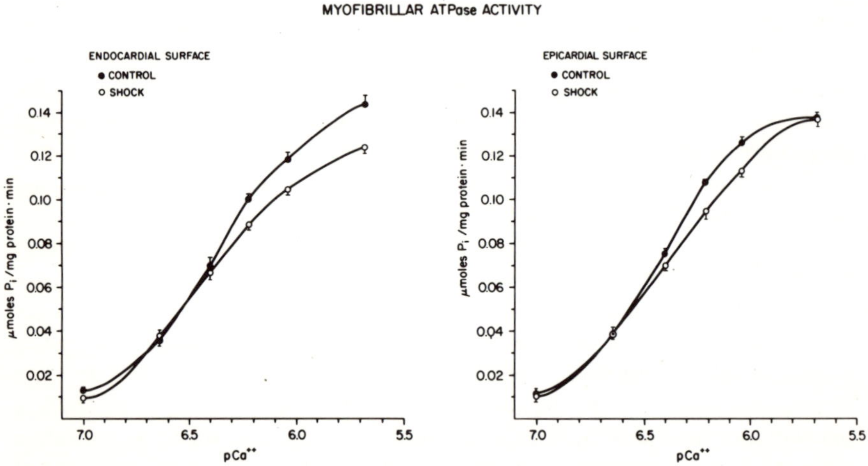

Fig. 9. Endocardial and epicardial myofibrillar ATPase activity expressed as μmoles of P_i per milligram protein-minute. From endotoxin-shocked hearts as compared to controls, over a range of intracellular free calcium concentrations. Each point represents the mean and each line represents one standard error of the mean.

no significant decrease in maximal ATPase activity, but there is a depression of ATPase activity on the ascending portion of the curve.

DISCUSSION

This study demonstrates several distinct and important differences between the in vitro and in vivo effect of endotoxin on the excitation–contraction coupling system of the myocardium. In vitro, endotoxin uncouples calcium transport from ATP hydrolysis in isolated cardiac sarcoplasmic reticulum. Endotoxin was found to be without effect on sarcoplasmic reticulum Ca^{2+}–ATPase, myofibrillar Ca^{2+}–ATPase and Na^{+}–K^{+} ATPase. In contrast, following the in vivo exposure of the myocardium to endotoxin during the shock state, there is a depression of both the sarcoplasmic reticulum and the myofibrils. This depression is greater at the subendocardial surface then at the subepicardial surface. Thus, it would appear that endotoxin, per se, is not inducing myocardial failure. If endotoxin were permeating the cell membrane and exerting an intracellular site of action, one should observe similar in vivo and in vitro results. Such is not the case. This observation also agrees with a similar conclusion of Dahalla et al [8] and Parker and Adams [9] who could find no direct effect of endotoxin on isolated cardiac muscle preparations. Thus, it would appear that endotoxin sets into motion a cascade of events, the final result being myocardial failure.

One may now ask, what is the pathogenic sequence of this indirect effect of endotoxin that results in myocardial failure? It has been shown by this laboratory [3] and others [10] that during the course of endotoxic and hemorrhagic shock, there is an increase in coronary vascular resistance and a decrease in coronary flow. In the working heart, this would place the more vulnerable subendocardium at greater jeopardy. This should result in subendocardial hemorrhagic necrosis and a depression of those subcellular organelles responsible for the initiation and termination of the excitation–contraction coupling process, an observation that this study confirms. Further support for this hypothesis comes from earlier work from this laboratory that demonstrated that an incorporation of an augmented venous return during the course of endotoxin shock increases coronary flow, decreases coronary vascular resistance, and protects the subendocardial sarcoplasmic reticulum and myofibrils [3]. Using a similar model of an augmented venous return, Beller-Todd et al [11] demonstrated an increase in survival rates and speculated that this may be due to a preservation of myocardial function.

In summary, it would appear that current evidence incriminates an indirect mechanism of action of endotoxin on myocardial function. It would appear that endotoxin-induced hypotension sets into motion a series of positive feedback mechanisms that results in a state of global ischemia secondary to an increase in coronary vascular resistance and a decrease in coronary flow. This hypothesis becomes more tenable when one considers that the subendocardial sarcoplasmic reticulum and myofibrillar function are more depressed than the epicardial population. The primary mediators of this circle of positive feedback still await definition.

ACKNOWLEDGMENTS

This work was performed during the tenure of M. L. Hess as an Established Investigator of the American Heart Association. It was supported by the Grant-in-Aid HL 24502-01 from the National Institutes of Health.

REFERENCES

1. Fabiato A, Fabiato F: Calcium and cardiac excitation–contraction coupling. Ann Rev Physiol 41:473, 1979.
2. Kleinman WM, Krause SM, Hess ML: Differential subendocardial perfusion and injury during the course of gram-negative endotoxemia. Advan Shock Res 4:139, 1980.
3. Bruni FD, Komwatana P, Soulsby ME, Hess ML: Endotoxin and myocardial failure: Role of the myofibril and venous return. Am J Physiol 225:H150, 1978.
4. Hess ML, Soulsby ME, Davis, JA, Briggs FN: The influence of venous return on cardiac mechanical and sarcoplasmic reticulum function during endotoxemia. Circ Shock 4:143, 1977.
5. Krause SM, Hess ML: Diphenhydramine protection of the failing myocardium during gram-negative endotoxemia. Circ Shock 6:75, 1979.
6. Soulsby ME, Ford GD, Davis JA, Wincklhofer K, Hess ML: The effect of Gram-negative endotoxin on calcium accumulation by subcellular fractions of vascular smooth muscle. Circ Shock 3:325, 1976.
7. Fabiato A, Fabiato F: Effects of pH on the myofilaments and sarcoplasmic reticulum of skinned cells from cardiac and skeletal muscles. J Physiol 276:233, 1978.
8. Beaumont L, Rawson L, Sulakhe PV, Dhalla NS: Calcium metabolism and contractility of isolated cardiac muscle from endotoxin-treated guinea pigs. In Dhalla NS (ed): Myocardial Biology: "Recent Advances in Studies on Cardiac Structure and Metabolism." Vol 4. Baltimore: University Park Press, 1972, p 305.
9. Parker JL, Adams HR: Myocardial effects of endotoxin shock: Characterization of an isolated heart muscle mode. Advan Shock Res 2:163, 1979.
10. Bohrs CT, Turbow ME, Kolmen SN, Traber DL: Coronary blood flow alterations in endotoxin shock and the response to dipyridamole. Circ Shock 3:281, 1976.
11. Rubanyi G, Kovach AGB, Koltay E, Nagy-Dora T, Balogh I, Somogyi E: Effect of hemorrhagic shock on the performance, 0_2 consumption and ultrastructure of isolated rat hearts. Circ Shock 7:59, 1980.
12. Beller-Todd B, Archer LT, Hinshaw LB: Recovery from endotoxin shock after extracorporal perfusion without anticoagulation. Circ Shock 6:261, 1979.

In Vitro Interactions of Endotoxin, Chromatin, DNA, and Steroid Hormone Receptors

Hassan Tavakoli and Robert J. Moon

Glucocorticoid hormones stimulate a wide variety of biological responses in animal tissues including protection against the lethal effects of bacterial endotoxin [1,2]. A large body of circumstantial evidence suggests that all steroid hormones act via basically the same mechanism. The hormones readily pass into the cytoplasm of cells [3,4] and bind to specific receptors [5, 6]. After a possible "activation" step [7–11], receptor–steroid complexes become concentrated in the nucleus [7,8,10,11]. The extent to which the receptor–steroid complexes bind to chromatin [10,12–14] or DNA [6,14–16] has not been definitively established. Whatever the precise mechanism(s), receptor–steroid complexes are thought to alter the transcription process resulting in increased levels of specific messenger RNAs [10,14]. The products of translation of these messenger RNAs leads to the observable biological effects [10,13,14], one of which is the induction of selected hepatic enzymes [17–28]. Bacterial endotoxin inhibits the induction of some, but not all, of these adaptive enzymes [2,29]. One possible molecular mechanism to explain endotoxin's ability to inhibit enzyme induction is that the lipopolysaccharide (LPS) directly interferes with glucocorticoid-receptor binding to cells or subcellular targets. Experiments described herein were designed to test the hypothesis that endotoxin can block steroid binding to hepatic cell chromatin and/or DNA. This hypothesis seems plausible since a previous report from this laboratory [30] has shown that endotoxin associates directly with hepatic parenchymal cells and their nuclei in vivo. To evaluate the biological significance of the observed in vitro interactions, most experiments described herein compare particular variables using both toxic and alkaline detoxified endotoxins.

MATERIALS AND METHODS

All procedures were performed at 0–4°C, unless otherwise specified.

Pathophysiological Effects of Endotoxins at the Cellular Level, pages 33–43

Animals

Female Spartan HA-ICR mice were used for enzyme studies. Female Charles River CD-1 mice were used for organ distribution of endotoxin. Food and water were available ad libitum.

Chemicals

Minimum essential medium (MEM) and fetal calf serum (FCS) were obtained from GIBCO (Grand Island, NY).

[1,2,^{3}H]Dexamethasone (20–40 Ci/mmol) was purchased from New England Nuclear (Boston, MA) or Amersham/Searle (Arlington Heights, IL). Micrococcal nuclease, calf thymus deoxyribonucleic acid, and hydrocortisone-21-sodium succinate were purchased from Sigma Chemical Company (St. Louis, MO).

Phosphoenolpyruvate Carboxykinase Assay

Phosphoenolpyruvate carboxykinase (PEPCK) activity was assayed by the method of Lane et al [31]. PEPCK was induced by injecting 5 mg of hydrocortisone-21-sodium succinate suspended in 0.5 ml of sterile saline subcutaneously into the interscapular region. Toxic or alkaline-treated detoxified endotoxin (50 μg/mouse), dissolved in saline, was injected intraperitoneally (ip) immediately after hydrocortisone. PEPCK activity was determined in liver homogenates 4 hr later.

Preparation of Radiolabeled Endotoxin and Pure Nuclei

Salmonella typhimurium endotoxin, extracted by the Westphal method, was purchased from Difco (Detroit, MI). Residual protein was removed by ultracentrifugation at 100,000g for 8 hr, following low-speed centrifugation at 8,000 rpm for 1 hr. ^{51}Cr-labeled endotoxin was prepared according to the method of Chedid et al [32]. Blobel and Potter's [33] method was used to prepare nuclei free of debris by sucrose gradient centrifugation.

Preparation of HTC-Cell Cytosol and Chromatin

Hepatoma tissue culture (HTC) cells were grown at 37°C in suspension cultures in MEM supplemented with 5% heat-inactivated fetal calf serum. Cells were harvested at 600g for 5 min, washed at least twice with an ice-cold buffer [0.04 M Tris (hydroxymethyl) aminomethane-hydrochloride, 0.1 M KCl, 0.001 M $MgCl_2$; pH 7.2], suspended in 1-vol homogenization buffer [0.02 M Tris (hydroxymethyl) methylglycine, 0.002 M $CaCl_2$, 0.001 M $MgC1_2$, 10^{-4} dithiothreitol, pH 7.4] and homogenized with a Dounce homogenizer. Homogenates were centrifuged at 800g for 10 min, and the supernatant solution (nonnuclear fraction) was separated from the pellet-containing nuclei which were reserved and used as described below. The nonnuclear fraction was further centrifuged at 35,000g for 15 min. Dexamethasone (5×10^{-8} M) was added to the supernatant and the mixture was centrifuged at 105,000g for 90 min. The supernatant (cytosol) was then incubated for at least 2 hr to allow steroid binding, followed by activation of the steroid receptors by supplementing the cytosol with 0.15 M NaCl and incubating at 20–25°C for 30 min [8]. Activated cytosol was finally passed through 20 ml of Sephadex G-25 equilibrated with homogenizing buffer containing 0.25 M sucrose to remove unbound steroid and salt. The macromolecular fraction containing activated hormone–receptor complexes (AHRC) was used immediately. AHRC had about 5 mg/ml of protein and were inactivated (IHRC) by incubation at 37°C for 1 hr.

The nuclear fraction was washed once in 10 vol of 0.01 M Tris, pH 8.0, at 0°C and allowed to swell for 2 hr in 3.5 vol of the same buffer. Nuclei were broken by sonication and centrifuged at 20,000g for 15 min. The supernatant contained chromatin and had about 1 mg/ml of DNA, as measured by the diphenylamine technique [34].

Binding of ^{51}Cr-Labeled Toxic and Alkaline-Treated Detoxified Endotoxin to Chromatin

Alkaline-treated endotoxin was prepared by dissolving 50 mg of *Salmonella typhimurium* endotoxin in 3 ml of 0.25 N NaOH. The solution was heated at 56°C for 60 min [35]. The modified endotoxin was precipitated with cold anhydrous ethanol, washed three times with ethanol, dialyzed against deionized distilled water, and then lyophilized. Both toxic and detoxified endotoxin were labeled with $Na_2{}^{51}CrO_4$ according to the method of Chedid et al [32].

An aliquot (20 μl) of chromatin was incubated for 90 min with 10 μl of ^{51}Cr-labeled toxic or alkaline-treated detoxified endotoxin. Chromatin precipitation was achieved by adding 2 ml of 5mM $MgCl_2$–0.01 M Tris buffer to the assay mixture. After at least 10 min, the mixture was centrifuged at 8,000g for 10 min. As a control, 10 μl of ^{51}Cr-labeled toxic or alkaline-treated detoxified endotoxin was mixed with 20 μl of homogenizing buffer and precipitated with $MgCl_2$ solution. The counts per minute of endotoxin precipitated by $MgCl_2$ in the control tubes was considered background and subtracted from the test values.

Binding of Hormone Receptor to Chromatin in the Presence of Toxic and Detoxified Endotoxin

Different concentrations of toxic or alkaline-treated detoxified endotoxin were mixed with 10 μl of chromatin. After making the volumes equal with homogenizing buffer, the endotoxin–chromatin mixtures or chromatin alone were incubated for 2.5 hr. To each tube, 0.4 ml of the macromolecular fraction of activated cytosol was added. After 90 min of incubation, 0.2 ml of the assay mixture was added to 1.8 ml of 5 mM $MgCl_2$–0.01 M Tris buffer. The tubes were mixed on a Vortex mixer and, after at least 10 min, centrifuged at 8,000g for 10 min. The pellet was digested in 0.4 ml of 1 M NaCl–0.5 M NaOH overnight at room temperature and counted in 3a70B cocktail (rpi). As a control, 0.2 ml of activated cytosol was added to 1.8 ml of $MgCl_2$ solution and processed as stated. Counts per minute of activated cytosol that were precipitated with $MgCl_2$ were subtracted from the test. The percentage of inhibition of binding of hormone receptor to chromatin in the presence of endotoxin was calculated as follows: 100-[cpm of hormone receptor bound to chromatin in the presence of endotoxin/cpm of hormone receptor bound to chromatin] × 100.

Binding of Endotoxin to DNA

DNA–cellulose (340 μg DNA/ml) was prepared according to the technique of Albert et al [36]. An aliquot of 0.1-ml or 0.3-ml DNA–cellulose or cellulose alone was centrifuged in 1.5-ml disposable conical tubes (Eppendorf) at 600g for 5 min. The supernatant was removed by aspiration. To each tube, 20 μl of labeled endotoxin was added and the total volume was made up to 0.6 ml with Tris–glycerol buffer (0.01 M Tris–10^{-4} M dithiothreitol–10% glycerol v/v). The tubes were capped, placed on a roller bottle apparatus at 10 rpm for 2.5 hr, and turned 180° every 30 min during the assay to help keep the cellulose resuspended. The reaction was stopped by centrifugation at 600g for 5 min. The supernatant was removed and the ^{51}Cr radioactivity was measured. The pellets were resuspended in Tris–glycerol buffer and counted.

Binding of Hormone Receptor to DNA in the Presence of Toxic Endotoxin

Aliquots of 0.1 ml of DNA–cellulose (34 μg) or cellulose alone were centrifuged in 1.5-ml disposable conical tubes at 600g for 5 min. The supernatant was discarded and the pellet was suspended in 0.6 ml of Tris–glycerol buffer. Different amounts of toxic endotoxin (100 or 300 μg) or Tris–glycerol buffer were added to DNA–cellulose and left on a roller bottle apparatus (10 rpm) for 1.5 hr and turned 180° every 30 min to keep the DNA–cellulose in suspension. After incubation, some of the tubes which contained 100 μg of toxic endotoxin were centrifuged, the supernatant was discarded, and the pellet was resuspended in 0.6 ml of Tris–glycerol buffer. The binding of AHRC to DNA in the presence or absence of excess toxic endotoxin was determined by raising the total volume of all the tubes to 0.8 ml by adding 0.2 ml of AHRC. The tubes were capped and left on a roller bottle apparatus for 2.5 hr. The reactions were stopped by centrifugation at 600g for 5 min. The pellets were resuspended in 0.2 ml of Tris–glycerol buffer for at least 1 hr at room temperature before being counted. The percentage of inhibition of binding of AHRC to DNA in the presence of toxic endotoxin was calculated as follows: 100 − (cpm of AHRC bound to

TABLE I. Hydrocortisone Induction of PEPCK in the Presence and Absence of Toxic and Detoxified Endotoxin

Experimental	PEPCK activity
Normal	19.7 ± 1.4[a]
Cortisone	34.2 ± 3.7[b]
Toxic endotoxin + cortisone	20.8 ± 2
Detoxified endotoxin + cortisone	34.9 ± 3

[a]Mean ± standard error of the mean obtained from six separate experimental determinations.
[b]Significantly different from toxic endotoxin + cortisone ($P < 0.05$).

DNA–cellulose which was incubated with endotoxin-cpm of AHRC bound to cellulose)/(cpm of AHRC bound to DNA cellulose-cpm of AHRC bound to cellulose) × 100.

RESULTS

Cortisone Induction of PEPCK in Normal and Endotoxin-Poisoned Mice

The effect of toxic and detoxified endotoxin on the in vivo induction of PEPCK by cortisone is presented in Table I. Cortisone administration significantly increased the level of PEPCK in the liver 4 hr after injection, and toxic endotoxin inhibited this induction. When endotoxin was detoxified by treating with 0.25 N NaOH, the ability of endotoxin to inhibit the induction of PEPCK in whole animals was prevented.

Hepatic Association and Subcellular Distribution of Toxic and Detoxified Endotoxin

Table II shows the extent of binding of intravenously (iv) injected toxic and detoxified endotoxin to liver and hepatic nuclei. Experimentally, 200 μg (ca 120,000 cpm) of ^{51}Cr-labeled toxic endotoxin was injected into mice. After 5 hr about 50% of the label (65,643 cpm) was associated with the liver. Of this number 7,462 cpm could be recovered from washed crude nuclei, and 366 cpm from pure nuclei. Five hours after injection of 180 μg (45,000 cpm) of detoxified endotoxin, hepatic association was significantly lower than with toxic endotoxin. About 50% of detoxified endotoxin remained in the circulation and carcass (data not shown) on a percentage basis. No significant difference in hepatic nuclear association of toxic and 0.25 N NaOH-treated endotoxin was observed.

Ability of Toxic and Detoxified Endotoxin to Bind to Chromatin

Table III shows the ability of ^{51}Cr-labeled toxic and detoxified endotoxin to bind to sonicated chromatin. When 20 μl of chromatin (ca 20 μg of DNA) was

TABLE II. Hepatic Association and Subcellular Distribution of Toxic and Detoxified Endotoxin 5 hr after IV Injection into Mice

Experimental	Toxic endotoxin[a] (cpm)	Detoxified endotoxin[b] (cpm)
Liver homogenate	65,643 ± 4787[c]	10,878 ± 674[d]
Crude nuclear fraction	7,462 ± 775	1,140 ± 50
Purified nuclei	366 ± 27	139 ± 22

[a]200 μg (120,000 cpm)/mouse intravenously.
[b]180 μg (45,000 cpm)/mouse intravenously.
[c]Mean ± standard error of the mean obtained from six separate experimental determinations.
[d]Mean ± standard error of the mean obtained from four separate experimental determinations.

TABLE III. Binding of ^{51}Cr-Labeled Toxic and Detoxified Endotoxin to Chromatin

Experimental	Added (μg)	Bound (μg)	Binding (%)
Toxic endotoxin	51.4	1.0	1.9
Detoxified endotoxin	46.0	0.24	0.52

incubated at 0°C with 51.4 μg of toxic endotoxin, 1.9% of the counts were bound after 90 min. Detoxification of endotoxin lowered chromatin binding approximately fourfold.

Effect of Graded Quantities of Toxic and Detoxified Endotoxin on Binding of AHRC to Chromatin

Graded quantities of unlabeled toxic endotoxin could inhibit binding of AHRC to a constant amount of sonicated chromatin (10 μl). In all experiments sonicated chromatin was used. No significant binding of AHRC to nuclease-extracted chromatin [37] was observed (data not shown).

Table IV shows that, as the concentration of toxic unlabeled endotoxin increased from 5 to 500 μg, the percentage of inhibition of binding of AHRC to chromatin increased up to a maximum of slightly higher than 50% with 500 μg of endotoxin. There was about 345 cpm of AHRC bound per 10 μg of DNA in chromatin. When AHRC was inactivated at 37°C for 1 hr (IHRC), no binding of IHRC to chromatin was observed. By contrast, alkaline-treated endotoxin exerted more of an inhibitory effect on the binding of AHRC than the toxic endotoxin. In the presence of 500 μg of alkaline-treated endotoxin, there was essentially 100% inhibition in the binding of AHRC to chromatin (Table IV).

TABLE IV. Inhibition of AHRC Binding to Chromatin in the Presence of Toxic and Detoxified Endotoxin

	Endotoxin (μg)							HRC bound to chromatin (cpm)		
Type of endotoxin	0	5	30	100	200	300	500	AHRC[a]	IHRC[b]	Blank[c]
Toxic endotoxin	0[d]	4.1	7.3	10.8	18.3	33.0	51.2	798	403	453
Detoxified endotoxin	0	22.5	34.8	67.0	—	97.4	100	891	390	558

[a]AHRC, activated hormone-receptor complex.
[b]IHRC, inactivated hormone-receptor complex.
[c]Blank, $MgCl_2$ + AHRC.
[d]Percentage of inhibition of binding compared to AHRC binding to chromatin.

TABLE V. Binding of ^{51}Cr-Labeled Endotoxin to Cellulose or DNA-Cellulose

Experimental	Bound (cpm)
Cellulose (0.1 ml)	13,605
Cellulose (0.3 ml)	33,531
34 μg DNA-cellulose (0.1 ml)	10,358
100 μg DNA-cellulose (0.3 ml)	26,516

These results do not establish whether the inhibition of binding is a result of direct interaction of endotoxin with chromatin (cf Table III) or whether free endotoxin remaining in solution interacts directly with AHRC, discouraging binding.

Binding of Toxic Endotoxin to DNA and Its Effect on the Binding of AHRC to DNA

Two different concentrations of DNA (34 and 100 μg) were used to investigate whether or not toxic endotoxin was able to bind directly to DNA conjugated to cellulose (DNA–cellulose). Data in Table V shows that toxic endotoxin does not bind directly to DNA. When toxic endotoxin (88 μg) was added to either 0.1 or 0.3 ml of cellulose alone, 13,605 and 33,531 cpm of sedimentable label were observed, respectively. When DNA was conjugated to the cellulose, there was actually less binding than to cellulose alone, suggesting that endotoxin does not bind to DNA.

Table VI shows the effect of two different concentrations of endotoxin on the binding of AHRC to 34 μg of DNA in the form of DNA–cellulose. There was about 9.5% inhibition in the binding in the presence of 100 μg of endotoxin and

TABLE VI. Binding of AHRC to DNA-Cellulose in the Presence and Absence of Endotoxin

Experimental	AHRC bound (cpm)	Inhibition (%)
Cellulose	730	—
DNA-cellulose	3,236	—
DNA-cellulose + 100 μg of endotoxin	2,999	9.5
DNA-cellulose + 300 μg of endotoxin	2,871	14.6
Endotoxin-exposed DNA-cellulose	3,300	—

14.6% inhibition in the presence of 300 μg of endotoxin. When DNA–cellulose was centrifuged after incubation with 100 μg of endotoxin and resuspended in Tris–glycerol buffer, there was no inhibition of AHRC binding (line 5).

DISCUSSION

Despite the extensive effort devoted to understanding the pathophysiology of bacterial endotoxin, so far there are no conclusive data as to whether endotoxin exerts its primary toxic effects directly on cells or through release of soluble mediator(s). In the present study we have explored an in vitro model which could conceivably yield insight into the nature of metabolic changes observed in endotoxemia and the protective mechanism of glucocorticoids against endotoxin lethality. In this study we have used a line of glucocorticoid-responsive, rat hepatoma (HTC) cells. This cell line is particularly well suited for such a study because at least five steroid-induced responses have been defined [38] in these cells.

Since glucocorticoids are believed to increase enzyme synthesis by elevating the production of specific messenger RNAs [10,39], it might be postulated that endotoxin suppresses the hormonal induction of hepatic enzymes through an effect on RNA synthesis. However, when a sublethal dose of toxic endotoxin is given to adrenalectomized mice, total net RNA and protein synthesis in the liver increases [40]. If endotoxin acts by suppressing RNA synthesis, there must be a high degree of selectivity in the action of endotoxin. If hepatic parenchymal cells are among the direct targets of endotoxin action, endotoxin might be able to inhibit the hormonal induction of hepatic enzymes by competing with the hormone for possible binding sites on or within the target cells. Our results tentatively suggest this possibility in that endotoxin can interfere with AHRC binding to chromatin in vitro.

The whole cell uptake of [^{3}H] dexamethasone is not altered in the presence of 350 μg of endotoxin (results not presented). It, therefore, seems unlikely that endotoxin interferes with the cell uptake of the hormone. The fact that endotoxin binds to isolated chromatin in vitro (Table III) suggests that the association of endotoxin to chromatin may block the ability of hormone receptors to bind to chromatin. While this tends to support the idea that endotoxin attaches to nuclear chromatin and inhibits AHRC binding, two lines of data suggest caution must be exercised in drawing this conclusion. First, endotoxin seems to bind to chromatin by interaction with proteins rather than DNA. This is suggested indirectly by the observation that endotoxin cannot bind to calf thymus DNA associated with cellulose (Table V). Second, AHRC can bind to DNA (Table VI) and sonicated chromatin (Table IV), but it does not bind to deoxyribonuclease-treated nuclei [15] or nuclease-extracted chromatin, which might have lost between 25 and 50% of its DNA [37]. If the AHRC is binding to DNA and the endotoxin is binding to protein, a direct inhibition would not seem likely. Instead, some stereochemical alteration would have to be invoked as an explanation. Even though our data do not offer conclusive evidence, what would appear most likely is that the endotoxin interacts directly with the AHRC, blocking its ability to bind to DNA. This explanation is most consistent with our data, since the presence of endotoxin in solution inhibits, in a dose-dependent manner, the binding of AHRC to both chromatin (Table IV) and DNA–cellulose (Table VI). It is possible that endotoxin acts like the inhibitory material which has been detected in the cytosol of HTC cells and rat uteri [14,41]. This inhibitor interacts with the receptor–steroid complex [14] and prevents its binding to the acceptor sites in the nuclei.

The biological significance of the ability of endotoxin to block AHRC binding is further complicated by the need for relatively high concentrations of toxic endotoxin and by the observation that biologically detoxified endotoxin inhibits binding even more efficiently than toxic endotoxin (Table IV). These observations and the fact that bacterial endotoxin cannot inhibit the induction of the hepatic enzyme phosphoenolpyruvate carboxykinase by cortisol in vitro [42] probably indicate that endogenous factors [2,29] or a host-modified form of endotoxin is more likely to be involved in the inhibition of PEPCK induction than is a direct toxicity of endotoxin on hepatic parenchymal cells.

ACKNOWLEDGMENT

This study was supported by a grant from the Michigan Heart Association.

REFERENCES

1. Berry LJ, Smythe DS: Effects of bacterial endotoxins on metabolism. VII. Enzyme induction and cortisone protection. J Exp Med 120:721, 1964.
2. Berry LJ, Smythe DS, Colwell LS: Inhibition of inducible liver enzymes by endotoxin and actinomycin D. J Bacteriol 92:107, 1966.

3. Baxter JD, Tomkins GM: The relationship between glucocorticoid binding and tyrosine amino transferase induction in hepatoma tissue culture cells. Proc Nat Acad Sci USA 65:709, 1970.
4. Levinson BB, Baxter JD, Rousseau GG, Tomkins GM: Cellular site of glucocorticoid-receptor complex formation. Science 175:189, 1972.
5. Baxter JD, Tomkins GM: Specific cytoplasmic glucocorticoid hormone receptors in hepatoma tissue culture cells. Proc Nat Acad Sci USA 68:932, 1971.
6. Rousseau GG, Baxter JD, Higgins SJ, Tomkins GM: Steroid-induced nuclear binding of glucocorticoid receptors in intact hepatoma cells. J Mol Biol 67:99, 1972.
7. Buller RE, Toft DO, Schrader WT, O'Malley BW: Progesterone-binding components of chick oviduct. VIII. Receptor activation and hormone-dependent binding to purified nuclei. J Biol Chem 250:801, 1975.
8. Higgins SJ, Rousseau GG, Baxter JD, Tomkins GM: Early events in glucocorticoid: Activation of steroid receptor and its subsequent specific nuclear binding studied in cell free system. J Biol Chem 248:5866, 1973.
9. Kalimi M, Colman P, Feigelson P: The "activated" hepatic glucocorticoid–receptor complex: Its generation and properties. J Biol Chem 250:1080, 1975.
10. O'Malley BW, Means AR: Female steroid hormones and target cell nuclei. The effects of steroid hormones on target cell nuclei are of major importance in the induction of new cell function. Science 183:610, 1974.
11. Rousseau GG, Baxter JD, Higgins SJ, Tomkins GM: Steroid-induced nuclear binding of glucocorticoid receptors in intact hepatoma cells. J Mol Biol 79:539, 1973.
12. Beato M, Kalimi M, Konstam M, Feigelson P: Interaction of glucocorticoids with rat liver nuclei. II. Studies on the nature of the cytosol transfer factor and the nuclear acceptor site. Biochemistry 12:3372, 1973.
13. Marver D, Goodman D, Edelman IS: Relationships between renal cytoplasmic and nuclear aldosterone-receptors. Kidney Int 1:210, 1972.
14. Simons SS Jr, Martiney HM, Garcea RL, Baxter JD, Tomkins GM: Interaction of glucocorticoid receptor–steroid complexes with acceptor sites. J Biol Chem 251:334, 1976.
15. Baxter JD, Rousseau GG, Benson MC, Garcea RL, Ito J, Tomkins GM: Role of DNA and specific cytoplasmic receptors in glucocorticoid action. Proc Nat Acad Sci USA 69:1892, 1972.
16. Yamamoto KR, Stampfer MR, Tomkins GM: Receptors from glucocorticoid-sensitive lymphoma cells and two classes of insensitive clones: Physical and DNA-binding properties. Proc Nat Acad Sci USA 71:3901, 1974.
17. Feigelson P, Dashman T, Margolis F: The half-life time of induced tryptophan peroxidase in vivo. Arch Biochem Biophys 85:478, 1959.
18. Feigelson P, Greengard O: Regulation of liver tryptophane pyrrolase activity. J Biol Chem 237:1908, 1962.
19. Kenney FT, Flora RM: Induction of tyrosine α-ketoglutarate transaminase in rat liver. I. Hormonal nature. J Biol Chem 236:2699, 1961.
20. Knox WE, Auerbach VH: The hormonal control of tryptophan peroxidase in the rat. J Biol Chem 214:307, 1955.
21. Lin ECC, Knox WE: Adaptation of rat liver tyrosine-α-ketoglutarate transaminase. Biochim Biophys Acta 26:85, 1957.
22. Nickol JM, Lee KL, Kenney, FT: Changes in hepatic levels of tyrosine aminotransferase messenger RNA during induction by hydrocortisone. J Biol Chem 253:4009, 1978.
23. Nordlie RC, Lardy HA: Mammalian liver phosphoenolpyruvate carboxykinase activities. J Biol Chem 238:2259, 1963.
24. Nordlie RC, Varrichio FE, Holton DE: Effects of altered hormonal states and fasting on rat liver mitochondrial phosphoenolpyruvate carboxykinase levels. Biochim Biophys Acta 79:214, 1965.

25. Shrago E, Lardy HA, Nordlie RC, Foster DO: Metabolic and hormonal control of phosphoenol pyruvate carboxykinase and malic enzyme in rat liver. J Biol Chem 238:3188, 1963.
26. Thompson EB, Tomkins GM, Curran JF: Induction of tyrosine α-ketoglutarate transaminase by steroid hormones in a newly established tissue culture cell line. Proc Nat Acad Sci USA 56:296, 1966.
27. Weber G: Study and evaluation of regulation of enzyme activity and synthesis in mammalian liver. Advan Enzymol Reg 1:1, 1963.
28. Weber G, Srivastova SK, Singhal RL: Role of enzymes in homeostasis. VII. Early effects of corticosteroid hormones on hepatic glyconeogenic enzymes, ribonucleic acid metabolism and amino acid level. J Biol Chem 240:750, 1965.
29. Berry LJ, Smythe DS, Colwell LS: Inhibition of inducible liver enzymes by endotoxin and actinomycin D. J Bacteriol 92:107, 1966.
30. Zlydaszyk JC, Moon RJ: Fate of ^{51}Cr-labeled lipopolysaccharide in tissue culture cells and livers of normal mice. Infect Immunol 14:100, 1976.
31. Lane MD, Chang HC, Miller RS: Phosphoenolpyruvate carboxykinase from pig liver mitochondria. Methods Enzymol 13:270, 1969.
32. Chedid L, Skarnes RC, Parant M: Characterization of ^{51}Cr-labeled endotoxin and its identification in plasma and urine after parenteral administration. J Exp Med 117:561, 1963.
33. Blobel G, Potter VR: Nuclei from rat liver: Isolation method that combines purity with high yield. Science 154:1662, 1966.
34. Burton K: Determination of DNA concentration with diphenylamine. In Grossman L, Moldave K (eds): "Methods in Enzymology, Vol XII. Nuclei Acids, part B." New York: Academic Press, 1968, p 163.
35. Goodman GW, Sultzer BM: Mild alkaline hydrolysis of lipopolysaccharide endotoxin enhances its mitogenicity for murine B cells. Infect Immunol 17:205, 1977.
36. Alberts B, Herrick G: DNA–cellulose chromatography. Methods Enzymol 21:198, 1971.
37. Noll M, Thomas JO, Kornberg RD: Preparation of native chromatin and damage caused by shearing. Science 187:1203, 1975.
38. Simons SS Jr, Baxter JD, Garcea RL, Tomkins GM: Role of glucocorticoid receptors in steroid-induced protein synthesis. In Slavkin HC, Greulich RC (eds): "Extracellular Matrix Influences on Gene Expression." New York: Academic Press, 1975, p 37.
39. Mohla S, DeSombre ER, Jensen EV: Tissue specific stimulation of RNA synthesis by transformed extradiol-receptor complex. Biochem Biophys Res Commun 46:661, 1972.
40. Shtasel TF, Berry LJ: Effect of endotoxin and cortisone on synthesis of ribonucleic acid and protein in livers of mice. J Bacteriol 97:1018, 1969.
41. Chamness GC, Jennings AW, McGuire WL: Estrogen receptor binding to isolated nuclei. A nonsaturable process. Biochemistry 13:327, 1974.
42. Berry LJ, Goodrum KJ, Ford CW, Resnick IG, Shackleford GM: Partial characterization of glucocorticoid antagonizing factor in hepatoma cells. In Schlessinger D (ed): "Microbiology 1980." Washington DC: American Soc Microbiol, 1980, p. 77.

Membrane Effects of Endotoxins

Endotoxin-Initiated Membrane Changes in Rabbit Platelets

David C. Morrison, Zenaida G. Oades, and Delores Di Pietro

The participation of bacterial endotoxins in the pathogenesis of Gram-negative septic shock and disseminated intravascular coagulation is well established. In this respect, the administration of either endotoxins or their purified lipopolysaccharide (LPS) component into a variety of experimental animals faithfully reproduces many of the hematologic, hemodynamic, and metabolic changes characteristic of patients with Gram-negative sepsis. More recent studies have documented the critical importance of the lipid A region of the LPS molecule for the expression of biological activity.

In contrast to this well-defined relationship of the structure of endotoxin to its toxic properties is the uncertainty of precise mechanism(s) by which these potent bacterial products initiate these multiple host responses. Endotoxins, and in particular lipid A, have been clearly demonstrated to interact with a variety of host tissues and cells [1–3], and this has hampered efforts to define those potential targets which may be critical to the initiation of disease. Nevertheless, it would appear that not all interactions which endotoxins can initiate will be contributory factors to endotoxin shock [4, 5]. Further, recent experiments from our laboratory [6] have provided strong evidence to suggest that it is the capacity of endotoxins to interact directly and functionally with host cells and tissues, rather than their ability to localize in such tissues, which is the critical factor in determining host responses to endotoxins.

As the various contributors to this volume will clearly document, endotoxins will perturb virtually every cell with which they come in contact, presumably by interactions which are initiated at the cell membrane. This rather ubiquitous capacity of endotoxins (lipid A) to bind to mammalian cell membranes has led several investigators to look for specific endotoxin receptors on cell membranes [7–9]. The results of other investigations have demonstrated a rather generalized affinity of endotoxins for lipid bilayers, suggesting that nonspecific interactions may be of importance in initiating cell responses [10, 11]. As recently sum-

Pathophysiological Effects of Endotoxins at the Cellular Level, pages 47 – 64

marized by Morrison and Rudbach [12], there is currently no compelling evidence to support one or both of these types of interactions as being responsible for the multiple biological activities of lipid A. The reports by Coutinho and co-workers [9, 13, 14] have provided some immunologic evidence for specific endotoxin receptors; however, at least one study has failed to confirm these results [15]. Nevertheless, from the concept of potential therapeutic intervention in cases of clinical endotoxemia, the possibility of identification and subsequent neutralization of specific endotoxin receptors has considerable theoretical appeal.

Our own approach to deal with this question of how endotoxins interact with membranes was derived from an extensive series of experiments designed to examine in vitro the interaction of endotoxins with rabbit platelets. These studies have documented that complement-mediated rabbit platelet lysis initiated by endotoxins requires the participation of the alternative pathway of complement [16]. We have further shown that the mechanism involves binding of the endotoxin to the platelet membrane via its lipid A region [17]. The polysaccharide region of the endotoxin then provides the appropriate surface for the assembly of alternative complement pathway components and subsequent platelet lysis. In the course of these studies, we made the rather fortuitous observation that both isolated lipid A as well as LPS derived from rough (polysaccharide-deficient) mutants had the capacity, under the appropriate experimental conditions, to elicit a complement-independent secretory response. The available data suggest that important interactions between the LPS and the platelet membrane must occur to allow a subsequent secretory response. This manuscript characterizes this unique platelet secretory response to lipid A and discusses its potential relevance to lipid A-mediated membrane-perturbational events in general.

MATERIALS AND METHODS

Bacterial Endotoxins

Purified lipopolysaccharides from *Escherichia coli* 0111:B4 were prepared by phenol extraction and subsequent fractionation and purification by gel filtration chromatography as previously described [18]. This procedure results in two well-defined LPS components which differ markedly in their content of lipid A and polysaccharide. Extraction of LPS from the rough (galactose-epimerase-deficient) J-5 mutant of *E. coli* 0111:B4 and deep-rough *Salmonella minnesota* R595 mutant was performed exactly as described by Galanos et al [19]. These latter LPS preparations were routinely sonicated for several minutes just prior to use in experiments. Radioiodination of R595 LPS was performed exactly as described by Ulevitch [20] and was provided by Dr. Ulevitch (Scripps Clinic and Research Foundation, La Jolla, CA) for use in these experiments. Specific activity of the labeled LPS was approximately 7 μCi/μg of LPS.

Rabbit Platelet Preparation

Blood was collected from the central ear artery of healthy New Zealand white rabbits (3–4 kg) of both sexes (Rancho de Conejo, Vista, CA) into 1/6 vol of acid citrate dextrose. Following centrifugation, the platelet-rich plasma was incubated for 20 min at 37°C with 0.5 μCi/ml of [^{3}H]serotonin (5-OH-tryptamine, 20 mCi/mM, Amersham Corp., Arlington Heights, IL), exactly as previously described [16, 17]. The platelets were then washed several times, resuspended in Tris-buffered saline, pH 7.4, and maintained at room temperature until used for experiments.

Mouse Platelet Preparations

Blood was obtained from healthy 25- to 30-gm C3H/St (L. C. Strong Co., Del Mar, CA) and C3H/HeJ (Jackson Lab., Bar Harbor, ME) mice of both sexes by retro-orbital puncture and collected into heparinized capillary tubes. Platelets in platelet-rich plasma were labeled with [^{3}H]serotonin and then isolated exactly as described in our recent publication [21].

Platelet-Release Assays

The release of [^{3}H]serotonin was used as a measure of platelet-secretory responses [16]. Platelet lysis was assessed by release of lactic dehydrogenase [22]. Rabbit platelets were utilized at a concentration of 5×10^8/ml and mouse platelets at about 1/10 that concentration. We have detailed the exact protocols for these experiments in several earlier publications. For most experiments, the results represent the average of duplicate samples which routinely varied by less than 10%. All data represent the results of experiments performed at least three times.

Binding Experiments

The binding of LPS and divalent cations ($^{45}CaCl_2$, 10 mCi/mg and $^{133}BaCl_2$, 5 mCi/mg, Amersham Corp.) to preparations was assessed following sedimentation of reaction mixtures through a mixture of dibutylphthalate: dinonylphthalate (4:1). Briefly, 100 μl of platelet suspensions were layered onto 300 μl of the phthalate mixture and centrifuged for 2 min in a Beckman microfuge (Beckman Instruments, Fullerton, CA). The bottom tip of the microfuge tube was then cut off and counted. The ^{125}I-LPS was assayed directly in a γ-radiation spectrometer. The ^{45}Ca and ^{133}Ba were solubilized in 10 ml of a 1:1 mixture of Aquasol (NEN, Boston, MA): toluene and assessed in a liquid scintillation spectrometer.

Electron Microscopy

All preparations of platelets were prefixed in 1% paraformeldehyde containing 0.72% gluteraldehyde in 0.2 M cacodylate buffer, pH 7.3, for 5 min, pelleted by centrifugation, and washed twice in phosphate-buffered saline.

For transmission electron microscopy approximately 10^8 platelets were fixed as described above for 30 min at room temperature and postfixed in 2% OSO_4 for an additional 20 min. Following extensive washing, cells were treated with saturated thiocarbohydrazide for 5 min, dehydrated with 30% ethanol and stained with 2% uranylacetate. Following dehydration through a graded series of ethanols and propylene oxide, cells were embedded in epon. Sections were poststained with 2% uranylacetate and 0.01% lead citrate and examined with a Hitachi Model HU12-A electron microscope.

For scanning electron microscopy, approximately 2×10^8 platelets were spotted for 10 min on 5.0 μm nucleopore filters (Millipore Corp. Lexington, MA) previously saturated with 2.5% poly-L-lysine, and allowed to air dry. They were then processed as described by Kelly et al. [23], critically point dried, mounted on aluminum stubs, and sputter coated with 100 Å of gold palladium. Samples were examined in a Hitachi HS-500 scanning electron microscope at 15 kV with a 5.0-mm working distance at 28° tilt.

RESULTS

Response of Rabbit Platelets to R595 LPS

Our initial experiment established that the incubation of rabbit platelets at 37°C with preparations of *S. minnesota* R595 LPS in Tris-buffered saline, followed by the addition of normal rabbit plasma, resulted in a dose-dependent release of [^{3}H]serotonin. We then established that the plasma component required for the release response was heat stable and dialyzable. Subsequent experiments demonstrated that the addition of millimolar concentrations of Ca^{2+} would substitute for the plasma requirement. The data shown in Figure 1 indicate a typical response of rabbit platelets to R595 LPS following addition of $CaCl_2$. Of importance in this experiment is the fact that the platelet response to calcium requires a preincubation of the platelets with the LPS. Further, the magnitude of the response increases linearly over an approximate 60- to 90-min period. The lack of significant response in the absence of calcium addition indicates that, at the concentration of LPS used, there was no significant release of platelet constituents in the absence of calcium.

The Platelet Response Is a Calcium-Dependent Secretion

As the platelet response to calcium, leading to the release of granule-localized serotonin, may be accomplished either by a noncytolytic secretory mechanism or, alternatively, by a direct lytic event, it was important to establish which of

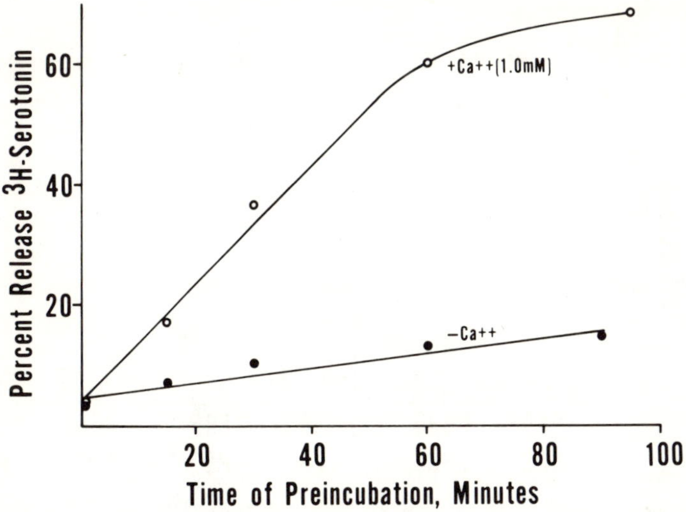

Fig. 1. Effect of time of preincubation of R595 LPS with washed rabbit platelets on calcium-induced secretion. Platelets, at a concentration of 5×10^8 per ml, were incubated at 37°C with 300 μg/ml of LPS from *S. minnesota* 595 for various lengths of time after which 1.0 mM CaCl was added and the incubation continued an additional 15 min. Platelets were then sedimented by centrifugation and the supernatants assayed for the release of [^{3}H]serotonin.

these mechanisms was responsible for the calcium-dependent release response described above. Further, in view of the importance of calcium in mediating secretory responses in general, and platelet secretion in particular, it was essential to establish that the observed response was specific for this divalent cation. Therefore, platelets were incubated with R595 LPS for 90 min (see Fig. 1) and then challenged with either Mg^{2+}, Ca^{2+}, or Ba^{2+}. Platelet responses were assessed for release of [^{3}H]serotonin (as a measure of release of granule constituents) and cytoplasmic lactic dehydrogenase (to estimate platelet lysis). The data shown in Figure 2 demonstrate that, of the three divalent ions tested, only Ca^{2+} was effective in initiating release of serotonin. The lack of release of LDH under any of the experimental conditions tested provides evidence that the Ca^{2+}-induced response is, indeed, secretory.

Although only Ca^{2+} was effective in mediating the secretory response to LPS-treated platelets, we queried whether the nonstimulatory ions might still be effective in competing with Ca^{2+}, and thus inhibiting secretion. The data shown in Figure 3 provide strong experimental support to suggest that this is not the case. Thus, the LPS-treated platelets challenged with 1 mM Ca^{2+} in the presence of 10-fold molar excess of either Mg^{2+} or Ba^{2+} were virtually indistinguishable

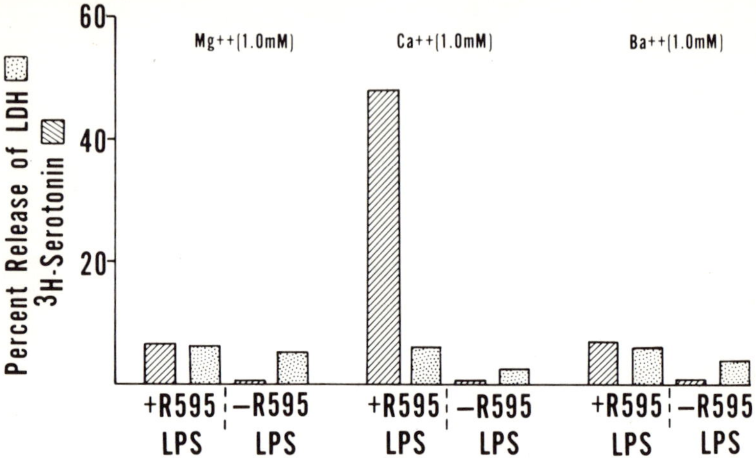

Fig. 2. Effect of divalent cations on R595 LPS-initiated rabbit platelet release response. Platelets were incubated at 37°C in either 400 μg/ml of *S. minnesota* R595 LPS or buffer. Then 1.0 mM $MgCl_2$, $CaCl_2$, or $BaCl_2$ was added and the incubation continued for 15 more min. After sedimentation of platelets, supernatants were assayed for [^{3}H]serotonin and lactic dehydrogenase (LDH).

from controls in their extent of secretion. As these experiments were performed following incubation with a suboptimal concentration of LPS, our data would suggest that the divalent cations which are not effective at initiating secretion are equally ineffective at blocking the Ca^{2+}-mediated secretions.

LPS Requirements for Platelet Secretion

As the LPS-induced secretion can be demonstrated with LPS from *S. minnesota* R595, a critical role for lipid A is clearly suggested. To examine whether the observed response is a general property of LPS preparations, we utilized several preparations of LPS which differ in their relative amounts of lipid A. As shown in Figure 4, both the R595 LPS as well as the LPS from the J5 (Ra) mutant of *E. coli* were effective at initiating calcium-dependent secretion. The polysaccharide-containing LPS preparations from *E. coli* 0111:B4, however, were either completely ineffective or effective only at inducing platelet lysis. We have also demonstrated (data not shown) that LPS from the semirough K-12 strain of *E. coli*, as well as isolated lipid A, will initiate calcium-dependent secretion. Thus, the observed response appears to be a general property of lipid A-rich preparations of LPS.

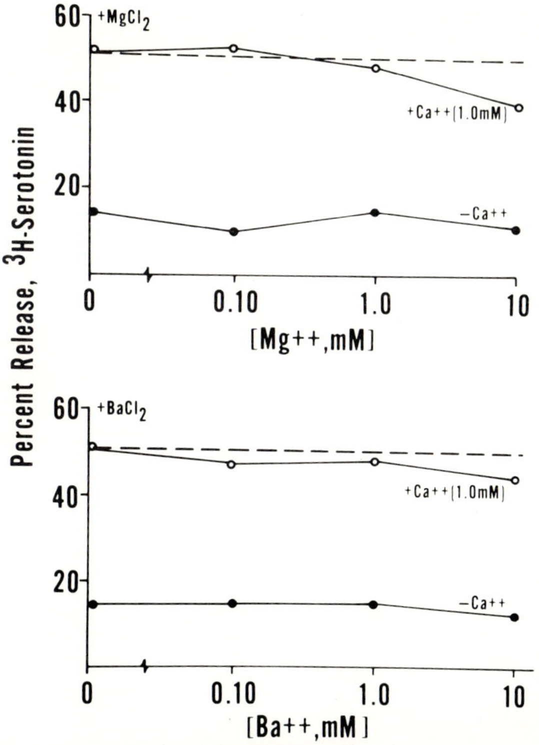

Fig. 3. Influence of divalent cations on R595 LPS-mediated calcium-initiated rabbit platelet secretion. Experimental conditions were as described in the legend to Figure 2, except that combinations of Ca^{2+} and or Mg^{2+} were added simultaneously to the LPS-treated platelets.

Parameters Regulating the Response to LPS

Our data, shown in Figure 1, indicate that the capacity of LPS-treated platelets to respond to Ca^{2+} is dependent upon the time of incubation of platelets with LPS prior to addition of Ca^{2+}. This might be explained by a time-dependent association of LPS with the platelet which eventually results in concentrations sufficient to modulate Ca^{2+} responsiveness. Alternatively, the capacity to recognize specifically Ca^{2+} could result from a critical time-dependent change in the nature of the LPS–platelet membrane relationship, which then allows effective recognition of Ca^{2+}. In order to help distinguish between these possibilities, experiments were performed in which the association of radiolabeled LPS with the platelets was examined relative to both the time and temperature of the

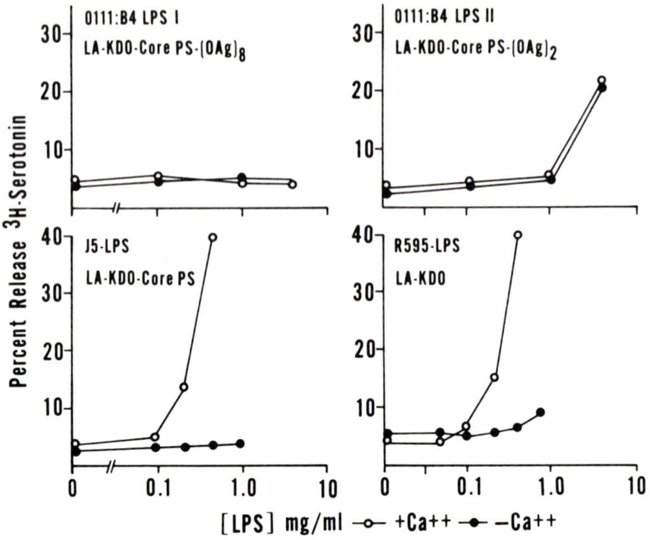

Fig. 4. Initiation of secretion from rabbit platelets by various LPS preparations. Platelets were incubated at 37°C with various concentrations of LPS from *S. minnesota* R595; *E. coli* J5 or *E. coli* 0111:B4 (fractions I and II) for 90 min. Then either calcium-free buffer or $CaCl_2$-containing buffer (final concentration 1.0 mM) was added to parallel tubes and release assessed as described in the legend to Figure 1.

LPS–platelet preincubation. Platelet secretory responses were then subsequently assessed by warming platelets to 37°C with Ca^{2+}-containing buffer. The results of this experiment are shown in Figure 5 and illustrate several important points. First, the association of LPS with the platelet (binding) is virtually instantaneous and is not statistically different with respect to either time or temperature. In contrast, the capacity of such LPS-treated platelets to secrete in response to Ca^{2+} is critically dependent upon both of these variables. Thus, whereas almost maximal secretion is obtained following incubation of platelets with LPS for 90 min at 37°C, this response is markedly reduced when the incubation is carried out at 33°C, and is abrogated completely at 23°C. We conclude from these experiments that the relevant interaction between the LPS and the platelet (membrane) occurs subsequent to a binding event and is both time and temperature dependent.

Morphological Alterations in LPS-Treated Platelets

To explore in more detail the effects which LPS induces in rabbit platelets, these cells were examined by electron microscopy for potential morphological changes following treatment with LPS. A portion of the results of this experiment

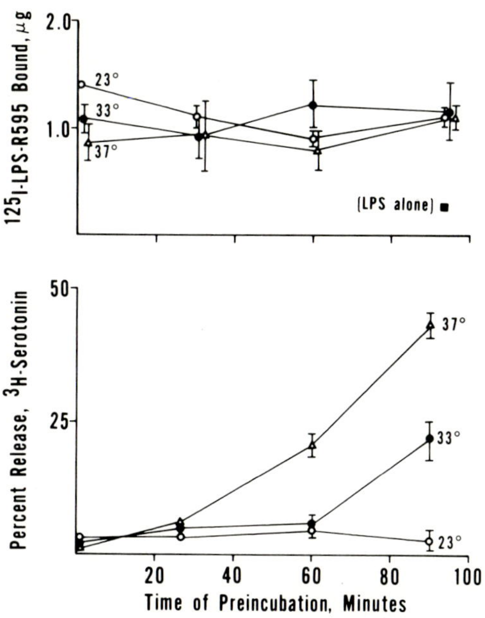

Fig. 5. Lack of correlation between binding of LPS and ability to initiate secretion. Parallel samples of platelets were incubated at 37, 33, or 23°C for various lengths of time with either radiolabeled *S. minnesota* R595 LPS (to assess binding) or unlabeled LPS (to measure release). The former platelets were sedimented through phthalate as described under Materials and Methods to measure binding of LPS to the platelets. To the latter cells was added 1.0 mM $CaCl_2$ and release of [^{3}H]serotonin determined after an additional 15-min incubation.

is shown in Figure 6. In contrast to control platelets, incubated for 60 min with buffer (Fig. 6a) which maintain a discoid-shaped appearance, the LPS-treated platelets demonstrate striking morphological changes, including the appearance of numerous finger-like projections and pseudopodia (Fig. 6b). That these morphological changes are associated with the capacity to respond to the addition of Ca^{2+} is suggested by the fact that platelets treated with LPS at 23°C for 60 min are indistinguishable by scanning electron microscopy from the control platelets shown in Figure 6a.

In order to provide additional morphological evidence for these rather striking platelet changes in response to LPS, we have also examined these LPS-treated cells by transmission electron microscopy. The electron micrographs of control (Fig. 6c) and LPS-treated platelets (Fig. 6d) essentially confirm the results described above using scanning electron microscopy. These combined data would thus support the concept of major alterations in the platelet ultrastructure following its treatment with LPS.

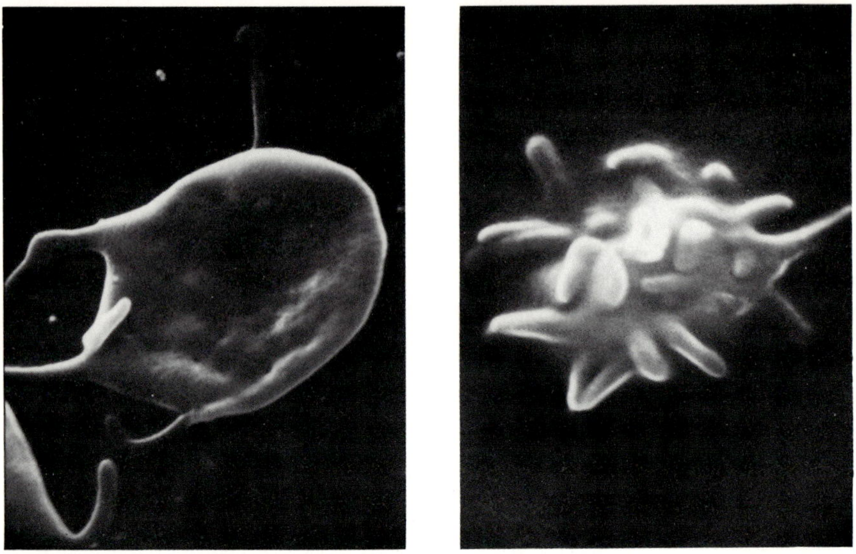

a b

Fig. 6. Electron micrographs of control and LPS-treated platelets. Suspensions of platelets were incubated for 90 min at 37°C with either 400 μg/ml of *S. minnesota* R595 LPS (Fig. 6a,c) or buffer (Fig. 6b,d). They were then examined by both scanning electron microscopy (Fig. 6a,b) or transmission electron microscopy (Fig. 6c,d).

Alteration in the Nature of the LPS Associated with the Platelet

Our combined data described above have indicated a time- and temperature-dependent alteration in the association of the LPS with the platelet which then allows the platelet–LPS to recognize specifically the presence of Ca^{2+} in the external environment. This specific recognition then results in the initiation of noncytolytic secretion of platelet granule constituents. These data would be consistent with the postulate of a direct alteration of the bound LPS either in or in the platelet membrane which then regulates Ca^{2+} binding and/or translocation. As a means of assessing this putative alteration, we have utilized several molecular probes to assess LPS-mediated changes in the platelet membrane.

The cationic polypeptide, polymyxin B, has been demonstrated by us [24] and by others [25] to bind with high affinity to the lipid A region of LPS and to abrogate a number of its biological activities (reviewed in [2, 3]). We have, therefore, assessed the effects of polymyxin B on the capacity of LPS to modulate platelet responses and to initiate the secretory response in LPS-pretreated pla-

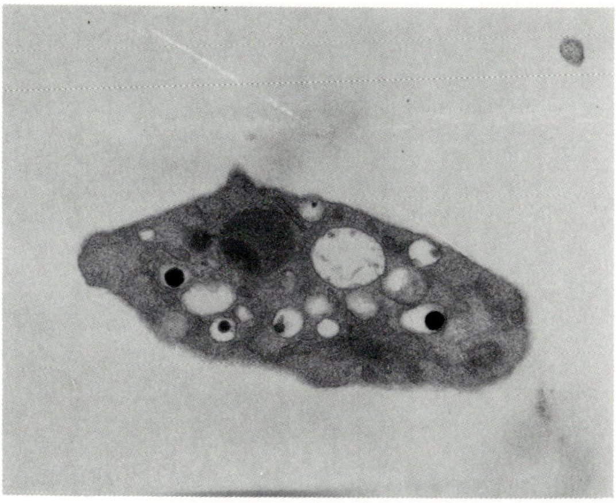

c

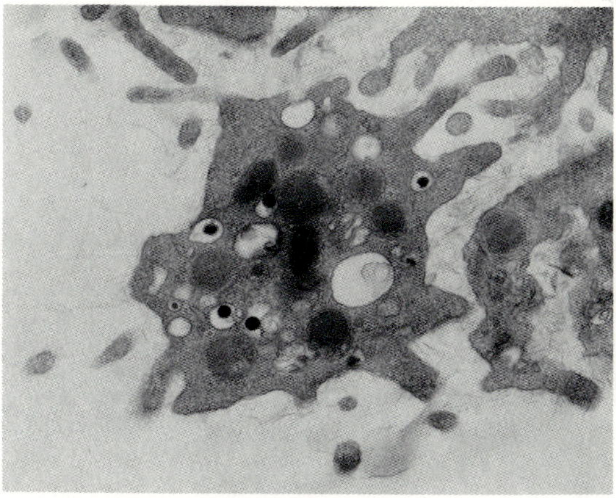

d

Figure 6

telets. As shown by the data in Figure 7, polymyxin B added to LPS-pretreated platelets at Time 0 has no direct effect on the release of granule constituents nor does it initiate platelet responses in the presence of LPS. Further, the addition of polymyxin B at Time 0 to LPS-treated platelets inhibits completely their capacity to respond subsequently to optimal concentrations of Ca^{2+} (data not shown).

A completely different situation is observed when 90-min LPS-pretreated platelets are subsequently challenged with polymyxin B. Also, as shown in Figure 7, the addition of polymyxin B to such cells results in the release of a

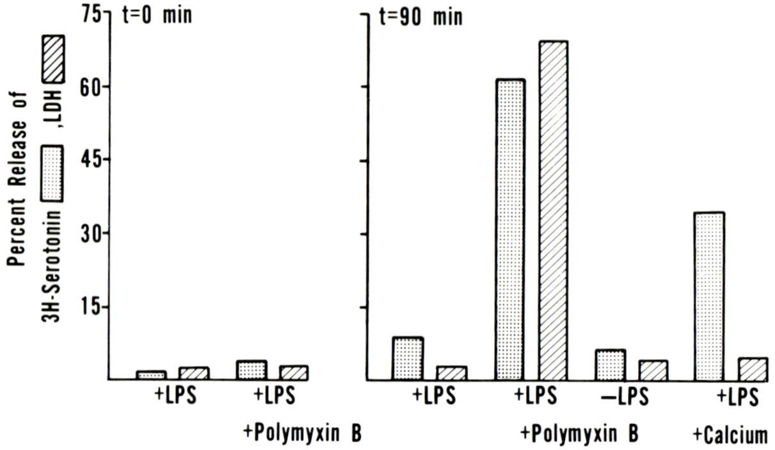

Fig. 7. Effect of polymyxin B on R595 LPS-treated platelets. Platelets were incubated for either 0 or 90 min at 37°C with 400 μg/ml of *S. minnesota* R595 LPS or buffer. Then either buffer, 1.0 mM $CaCl_2$, or 20 μg/ml of polymyxin B was added and the incubation was continued an additional 15 min. Release of [^{3}H]serotinin and LDH was assessed as described in the legend to Figure 2.

significant proportion of granule-localized serotonin, levels similar to those induced with millimolar Ca^{2+} (Fig. 7). In contrast to the Ca^{2+}-induced secretory response, however, the release triggered by polymyxin B appears to be cytotoxic as comparable levels of both [^{3}H]serotonin and cytoplasmic LDH are released in response to this stimulus. These data, considered with the results described, above would support the concept that the period of incubation of LPS with the platelets results in a modification of their association such that polymyxin B is recognized in a qualitatively and quantitatively distinct fashion.

It is, nevertheless, clear that the major relevant observation in the observed morphological and functional alteration of platelets by LPS is their capacity to recognize specifically the presence of Ca^{2+} in the initiation of a secretory response. We have, accordingly, assessed directly the capacity of LPS-treated platelets to bind and/or translocate $^{45}Ca^{2+}$. Since Ba^{2+} neither initiates secretion nor blocks the Ca^{2+}-induced secretion, we have utilized $^{133}Ba^{2+}$ as an appropriate nonstimulatory control. In addition, in separate but identical controls, we have assessed the capacity of these platelets to respond to Ca^{2+} with the release of [^{3}H]serotonin and to bind ^{125}I-LPS.

The results of these studies, representative of three experiments, are shown in Figure 8 and illustrate several major points. First, as demonstrated earlier (Fig. 5), the binding of LPS to the platelet is virtually instantaneous and does

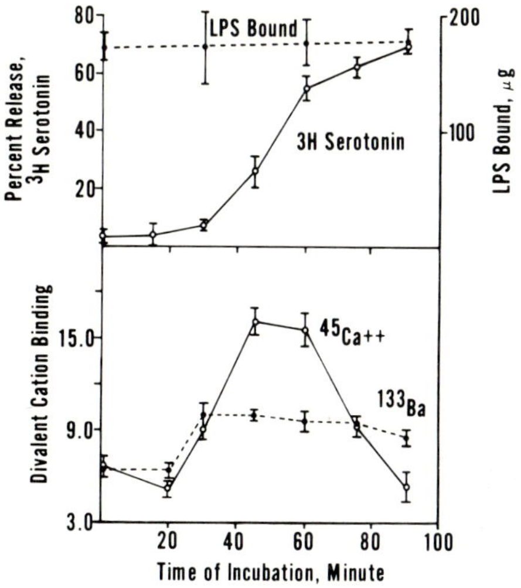

Fig. 8. Relationship of divalent cation binding to calcium-initiated secretion from LPS-treated platelets. Parallel samples of platelets were incubated with radiolabeled *S. minnesota* R595 LPS or unlabeled LPS (400 μg/ml) for various times. The platelets incubated with radiolabeled LPS were then sedimented immediately through phthalate to determine binding. To the platelets incubated with unlabeled LPS were added either $^{45}CaCl_2$ (10 mM, 1 μCi) or $^{133}BaCl_2$ (10 mM, 1 μCi) to measure binding of divalent cations or unlabeled $CaCl_2$ (10 mM) to assess release of [^{3}H]serotonin. After 10 min, the former two samples were sedimented through phthalate and the latter cells were pelleted by centrifugation. The various radiolabels were assayed as described under Materials and Methods.

not vary significantly with time. Second, there is a modest increase in the binding of $^{133}Ba^{2+}$ at 15–30 min after incubation with LPS; however, no statistically significant alterations in the binding of this nonstimulatory divalent cation can be detected subsequent to this early increase. In contrast, the association of $^{45}Ca^{2+}$ with the platelets following pretreatment with LPS increases dramatically with a time course coincident with its secretory response, assessed by the release of [^{3}H]serotonin. This is followed by a marked reduction in Ca^{2+} binding after extensive secretion has occurred during the latter time periods. Thus, whereas the binding of nonstimulatory divalent cations does not change radically following treatment of platelets with LPS, there is a marked temporal alteration in the capacity of Ca^{2+} to associate with these cells.

The Response of Platelets from the Endotoxin-Unresponsive C3H/HeJ Mouse to LPS

Our combined data have suggested a rather dynamic time- and temperature-dependent alteration in the nature of the association of LPS with the platelet membrane. Although not conclusive, these experiments suggest that LPS is perturbing the membrane in such a manner that it acquires the capacity to bind and/or translocate Ca^{2+}. Therefore, we queried whether such a response could be elicited in platelets from other species, in particular, platelets isolated from murine strains. These experiments were prompted by the now well-documented unresponsiveness of the C3H/HeJ strain of mice to respond to endotoxins (reviewed in [3]). This unresponsiveness is reflected not only in an absence of lethality in response to LPS but also in the inability of B lymphocytes, macrophages, and fibroblasts to respond to LPS. Thus, we argued, if our platelet model of calcium-dependent secretion following LPS stimulation accurately reflects a membrane-perturbational event and if such responses can be elicited in endotoxin-unresponsive mice, then the capacity of cell membranes in general from this strain of mouse could not be considered as defective in their capacity to be perturbed nonspecifically by LPS. This, then, would suggest that some biochemical event subsequent to, or in contrast to nonspecific membrane perturbation must be required for effective triggering of the lymphoid cells (eg, macrophages and lymphocytes).

We have, therefore, recently developed procedures for the isolation and measurement of secretory responses of murine platelets [21]. We then examined the capacity of the endotoxin-unresponsiveness C3H/HeJ as well as the histocompatible endotoxin-responsive C3H/St platelets to be stimulated by LPS. The results of this experiment are shown in Figure 9. Of importance, isolated platelets from both strains of mice responded equally to LPS as assessed by secretory responses following addition of Ca^{2+}. Although somewhat less dramatic than the responses elicited in preparations of rabbit platelets, they, nevertheless, represent significant temporally dependent increases above background values. These data thus provide some evidence to support the concept that the capacity of LPS to interact with cell membranes may not be the genetically controlled molecular defect which precludes their effective stimulation of C3H/HeJ lymphoid cells.

DISCUSSION

Our experiments have demonstrated a novel secretory response of rabbit and mouse platelets following their interaction with preparations of LPS containing high amounts of lipid A. The secretory response requires both a period of preincubation of LPS with the platelets and the subsequent addition of millimolar calcium. The magnitude of the secretory response is dependent upon both the time and temperature of the preincubation as well as the concentration of calcium. The observed response is specific for calcium and neither barium nor magnesium will substitute; nor will they block the calcium-induced secretion. Treatment of platelets with LPS results in rather profound morphological alterations in the platelet ultrastructure; however, integrity of the platelet membrane is maintained.

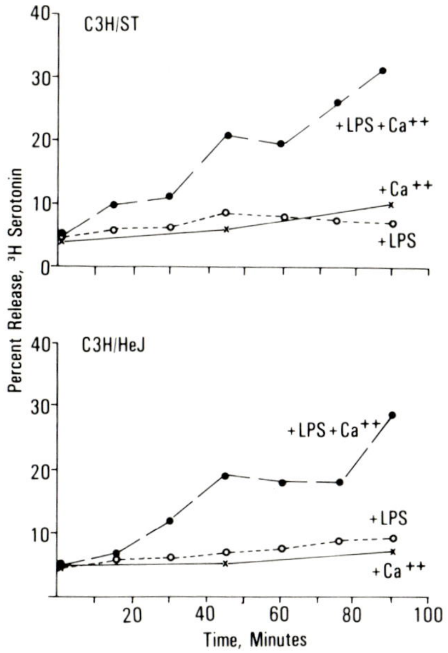

Fig. 9. Secretory responses of murine platelets to R595 LPS. Isolated platelets (~5 × 10^7/ml) from C3H/St mice (top panel) or C3H/HeJ mice (bottom panel) were incubated in Tris-buffered saline for various times with 300 μg/ml of *S. minnesota* R595 LPS. Then 10 mM $CaCl_2$ was added and the release of [^{3}H]serotonin was determined after an additional 10 min of incubation. Control platelets were incubated in the absence of LPS to which $CaCl_2$ was added ($+Ca^{2+}$) or in the presence of LPS to which calcium was not added (+LPS).

The acquisition of calcium responsiveness following incubation with LPS is accompanied by a preferential capacity of the platelets to bind and/or translocate calcium.

It has been recognized for a number of years that LPS has a high affinity for cations [26], in particular, divalent cations [27]. It is, therefore, of particular interest that only calcium will initiate a secretory response and that the capacity to respond requires a period of preincubation (processing). It is certainly plausible to argue that the addition of divalent cations to suspensions of LPS and platelets at early times either precipitates the LPS or prevents its effective interaction with the platelet membrane. However, whether the "processed" LPS still represents a selective target for calcium (but not other divalent cations) is more difficult to assess. The fact that these other divalent cations do not block the calcium-mediated response might argue against LPS being the binding site for calcium; however, additional experiments, currently in progress in our laboratory, will be required to distinguish these possibilities.

These combined data are consistent with, but not proof of, a two-step interaction of LPS with the platelet (membrane). The first may be considered a rapid time- and temperature-independent binding of LPS to the platelet. Subsequent to this is a second event perhaps involving the actual insertion of the LPS (or its subunits) into the platelet membrane, an event which would be expected to be dependent upon both time and temperature. This insertion event, as suggested above, then either allows the LPS to bind/translocate specifically calcium or makes accessible specific calcium sites on the platelet membrane required for secretion. It is, in fact, attractive to consider the "processed" LPS as acting functionally as an ionophore in this system. However, the predominant evidence from other experimental systems would not support ionophore-like activity of LPS as a general mechanism for LPS-initiated cell responses [12].

We consider it to be of particular significance that LPS-initiated calcium-dependent platelet responses can be demonstrated in the C3H/HeJ endotoxin-unresponsive mouse. These results may be contrasted with those demonstrating a total absence of response of C3H/HeJ macrophages, lymphocytes, and fibroblasts to the stimulatory activity of protein-free LPS [3]. As we have postulated that the platelet–LPS response is most probably a nonspecific perturbational event of the platelet membrane, these data would, therefore, suggest that the membranes of cells derived from the C3H/HeJ may not be defective in their capacity to interact with the lipid A of LPS. Of interest in this respect is the recently published report of Dumont [28] demonstrating that LPS will enhance the concanavalin A-initiated T lymphocyte mitogenic response equally well in C3H responder and nonresponder mice. If one were to argue that this enhancement effect were also nonspecific, these results would be consistent with those described in this manuscript. The extension of these observations would suggest that binding and/or nonspecific membrane-perturbational events initiated by LPS are not sufficient to result in cell activation, but that specific recognition by lipid A of high-affinity binding sites (endotoxin receptors) is also required. The recently published experiments of Forni and Coutinho [9, 13, 14] provide immunologic evidence for such a receptor, which is not present on lymphoid cells from the C3H/HeJ (or the endotoxin-unresponsive C57Bl/10SnCn) mouse. The functional identification and characterization of this endotoxin receptor will prove invaluable in defining the cellular targets of endotoxins in Gram-negative shock, as well as its precise biochemical mechanism of action.

ACKNOWLEDGMENTS

We thank Ms Julia Barber for outstanding secretarial assistance. This research was supported by NIH Grants 17225 and 17304 from the National Institute of Allergy and Infectious Disease.

REFERENCES

1. Kass EH, Wolff SM (eds): "Bacterial Lipopolysaccharides." Chicago: University of Chicago Press, 1973, passim.
2. Morrison DC, Ulevitch RJ: A review—The interaction of bacterial endotoxins with cellular and humoral mediation systems. Am J Pathol 93:527, 1978.
3. Morrison DC, Ryan JL: A review—bacterial endotoxins and host immune function. In Dixon FJ, Kunkel HG (eds): "Advances in Immunology." Vol. 28, New York: Academic Press, 1979, p 293.
4. Ulevitch RJ, Cochrane CG, Henson PM, Morrison DC, Doe WF: Mediation systems in bacterial lipopolysaccharide-induced hypotension and disseminated intravascular coagulation. I. The role of complement. J Exp Med 142:1570, 1975.
5. Curry BJ, Morrison DC: Role of complement in endotoxin-initiated lethality in mice. Immunopharmacology 1:125, 1979.
6. Musson RA, Morrison DC, Ulevitch RJ: Distribution of lipopolysaccharide in LPS-responsive and LPS-unresponsive mice. Infect Immunol 21:448, 1978.
7. Springer GF, Adye JC: Endotoxin-binding substances from human leukocytes and platelets. Infect Immunol 12:978, 1975.
8. Springer GF, Adye JC, Bezkorovainy A, Jirgensons B: Properties and activity of the lipopolysaccharide receptor from human erythrocytes. Biochemistry 13:1379, 1974.
9. Forni L, Coutinho A: An antiserum which recognizes lipopolysaccharides reactive B cells in the mouse. Eur J Immunol 8:56, 1978.
10. Rothfield L, Horne RW: Reassociation of purified lipopolysaccharide and phospholipid of the bacterial cell envelope: Electron microscopic and monolayer studies. J Bacteriol 93:1705, 1967.
11. Shands JW Jr: Affinity of endotoxin for membranes. In Kass EH, Wolff SM (eds): "Bacterial Lipopolysaccharides." Chicago: University of Chicago Press, 1973, p 189.
12. Morrison DC, Rudbach JA: Endotoxin–cell membrane interactions leading to transmembrane signals. In: Mandy WJ, Inman FP (eds): "Contemporary Topics in Molecular Immunology," Vol. 8, New York: Plenum Press, 1980 (in press).
13. Forni L, Coutinho A: Receptor interactions on the membrane of resting and activated B cells. Nature (London) 273:304, 1978.
14. Coutinho A, Forni L, Watanabe T: Genetic and functional characteristics of an antiserum to the lipid A specific triggering receptor on murine B lymphocytes. Eur J Immunol 8:63, 1978.
15. Watson J, Kelly K, Whitlock C: Genetic control of endotoxin sensitivity. In Schlessinger D (ed): "Microbiology—1980." Washington, DC: American Society for Microbiology, 1980, p 4.
16. Morrison DC, Kline LF, Oades ZG, Henson PM: Mechanisms of lipopolysaccharide-initiated rabbit platelet response. I. Alternative complement pathway dependence of the lytic response. Infect Immunol 20:744, 1978.
17. Morrison DC, Oades ZG: Mechanism of lipopolysaccharide-initiated rabbit platelet response. II. Evidence that lipid A is responsible for binding of lipopolysaccharide to the platelet. J Immunol 122:753, 1979.
18. Morrison DC, Leive L: Isolation and characterization of two fractions of lipopolysaccharide from *E. coli* 0111:B4. J Biol Chem 250:2911, 1975.
19. Galanos C, Luderitz O, Westphal O: A new method for the extraction of R lipopolysaccharides. Eur J Biochem 9:945, 1969.
20. Ulevitch RJ: The preparation and characterization of a radioiodinated bacterial lipopolysaccharide. Immunochemistry 15:157, 1978.

21. Oades ZG, Morrison DC: A simple and reproducible method to assess murine platelet responses in vitro. J Immunol Meth 34:79, 1980.
22. Henson PM: Release of vasoactive amines from rabbit platelets induced by antiplatelet antibody in the presence and absence of complement. J Immunol 104:924, 1970.
23. Kelley RO, Dekker RA, Bluemink JG: Ligand-mediated osmium binding: 1B. Application in coating biological specimens for SEM. J Ultrastr Res 45:254, 1973.
24. Morrison DC, Jacobs DM: Binding of polymyxin B to the lipid A portion of bacterial lipopolysaccharides. Immunochemistry 13:813, 1976.
25. Bader J, Teuber M: Action of polymyxin B on bacterial membranes. I. Binding to the O-antigenic lipopolysaccharide of *Salmonella typhimurium*. Z Naturforsch 28c:422, 1973.
26. Galanos C, Luderitz O: The role of the physical state of lipopolysaccharides in the interaction with complement. High molecular weight as prerequisite for the expression of anticomplementary activity. Eur J Biochem 65:403, 1976.
27. Galanos C, Luderitz O: Electrodrolysis of LPS and their conversion to uniform salt forms. Eur J Biochem 54:603, 1975.
28. Dumont F: Bacterial LPS enhances concanavalin A reactivity of thymocytes from the low LPS-responder C3H/HeJ mouse strain. Experientia 34:125, 1978.

Actions of Endotoxins on Excitation–Secretion Coupling at the Neural Membrane

Robert J. Person

Crude endotoxins (ETX) are high-molecular weight lipopolysaccharide–protein complexes derived from the cell wall of Gram-negative bacteria such as the genera *Escherichia, Shigella,* and *Salmonella*. Both ETX, the deproteinated lipopolysaccharide (LPS), and the parent live organisms have been used extensively to produce experimental shock as a model for clincially observed septic shock and for other forms of shock in which these endogenous bacteria presumably gain access to the systemic circulation. In sufficient concentrations ETX produces a lethal shock pattern characterized by an early phase of generalized vasoconstriction, hypotension, and the release of vasoactive agents, including catecholamines and prostaglandins [1–3]. The later phase of shock is marked by a continuous decline in normal physiological functions, inadequate perfusion of vital organs because of pooling of blood in flaccid capacitance vessels, and a significantly increased permeability of peripheral vessels.

Several lines of evidence converge on the excitable membrane as a possible primary target of these sepsis-producing toxins. Evidence based on systemic administration of ETX suggests that vascular smooth muscle undergoes a biphasic pattern of early constriction followed by an increasing loss of tone, suggesting a primary defect in excitation–contraction coupling. The early phase of shock is variously described as a stage of intense activation of the sympathoadrenal system [4] or a transient period of sympathetic hyperdischarge [3,6], thus suggesting defects in excitation–secretion coupling in autonomic neurons. There are significant changes in catecholamine metabolism indicating increased turnover [3,6] and a severe depletion of end organ catecholamines [6]. The latter study demonstrated a dose-dependent depletion of splenic and cardiac norepinephrine and adrenal epinephrine. Brain norepinephrine levels are also decreased and turnover is accelerated by either systemic or intracisternal injection of ETX with

Pathophysiological Effects of Endotoxins at the Cellular Level, pages 65 – 79

primary effects found in the hypothalamus and brain stem [6]. There is evidence that ETX may exert a highly specific effect on adenyl cyclase systems [7], which are now known to be important modulators of neurotransmitter and neurohumoral release. Taken together, these studies suggest that ETX increases the release and metabolism of neurotransmitters and hormones with a consequent depletion and eventual exhaustion of available stores. Further evidence of excitable membrane involvement, particularly the specialized membrane involved in excitation–secretion coupling, is found in studies showing that acute denervation of specific visceral organ beds prevents the usual pattern of vasoconstriction and accelerated catecholamine release in the denervated organ [8]. Further, close arterial injection exposing only splanchnic nerves to ETX is more effective in producing splanchnic area shock than the same dose administered systemically, and an ineffective systemic dose is highly toxic when administered locally [9,10].

Evidence of direct effects of ETX on the neural and muscle membrane has been reported by several investigators. Exposure of the isolated crayfish neuromuscular junction (at which the transmitter is GABA) to ETX concentrations reportedly equivalent to a rabbit LD_{50} (1–2 μg ETX/ml of bath) produces significant "facilitation" of evoked transmitter release and an alteration in spontaneous release characterized by an increased rate of multiquantal or "giant" miniature end plate potentials; both effects were proposed to be due to an increased calcium ion (Ca) conductance across the presynaptic membrane [11]. This proposal was later examined more directly by the same investigators and, in several preparations including smooth muscle, squid synapse, and frog atrium, effects were noted at similar ETX concentrations, which suggested that ETX causes an increase in transmembrane Ca currents under conditions in which the excitable membrane is activated [12].

Changes in Ca dynamics have also been observed in the isolated myocardium upon exposure to ETX. The toxin inhibits ATP-dependent Ca uptake in cardiac sarcoplasmic reticulum within several minutes (*Salmonella abortus equi* ETX, 25 μg/ml) [13], an effect which should result in a depression of myocardial contractility. Later work has, in fact, shown that isolated rat ventricle preparations suffer a decreased peak tension within 10 min after initial exposure to 0.1 μg/ml of ETX [Rudbach: ST-4 (0901)] resulting from a reduction in the duration of the active state [14]. The effect is reversed within 10 sec by a 50% increase in bath Ca concentration. This implies that ETX may interfere with the increase in Ca conductance across the sarcoplasmic reticulum required to link the membrane action potential with contractile molecules.

The implication of alterations in Ca dynamics induced by ETX is highly significant. Calcium ions are known to play a significant, if not universal, role in excitation–secretion and excitation–contraction coupling, not only at the presynaptic terminal and in muscle, but in hormone secretion generally [15]. It may well be that at least the early phase of shock induced by ETX is a direct result of changes in hormone secretion mediated by increases or decreases of Ca availability to the secretory or contractile machinery.

The Experimental Model

The amphibian, usually frog, neuromuscular junction is a useful preparation for the study of the interaction of endotoxins and excitation–secretion coupling. It is a large, highly efficient synapse at which intracellular recording of synaptic potentials can be accomplished with some facility, and its behavior under normal and a wide range of abnormal conditions is well described (cf, [16]). The application of standard intracellular recording techniques using glass capillary microelectrodes yields two types of synaptic potentials which result from the release of acetylcholine (ACh) through the presynaptic membrane. One type of synaptic potential, the end plate potential (EPP), is that which is usually effective in eliciting contraction of the muscle cell. Normally, it is 30–40 mV in amplitude and the secretion of the large amount of ACh required to produce the EPP is controlled by the entry of Ca from extracellular space during the opening of "slow" Ca channels in the presynaptic membrane. Opening of the Ca channels, which allows Ca to flow down a large electrochemical gradient into the cell interior, is initiated by depolarization of the presynaptic terminal during an action potential. Agents or conditions which antagonize the Ca channels, inhibit Ca entry, or ions such as magnesium (Mg) which compete with Ca for channel entry and are less effective in participating in the intracellular steps in excitation–secretion coupling, cause a decrease in the amount of depolarization-evoked transmitter release. This decrease is then reflected in a decreased amplitude of the postsynaptically recorded EPP. Conversely, agents which facilitate the transfer of Ca across the presynaptic membrane would increase EPP amplitude.

The second type of postsynaptically recorded potential is the miniature end plate potential (MEPP). It is a miniature EPP in all respects (0.5–1 mV in amphibia) save that it occurs spontaneously and presumably results from the randomly timed release of a quantum of transmitter, the quantum traditionally being identified as the contents of one synaptic vesicle. The rate of MEPP appearance, or MEPP frequency, is entirely controlled by presynaptic factors. The rate is proportional to presynaptic internal Ca concentration and agents or factors which increase or decrease free intracellular Ca, increase or decrease, respectively, MEPP frequency. Under most known conditions the only factors which will change the amplitude of MEPPs are factors which alter the sensitivity of the postsynaptic ACh receptors (eg, curare) or inhibit the hydrolytic breakdown of the ACh molecule (ie, anticholinesterases).

Experiments testing the effects of various toxins on spontaneous or evoked transmitter release are generally performed in vitro so that one or more neuromuscular end plates can be impaled with microelectrodes. MEPPs are recorded as they occur spontaneously and EPPs are recorded after a suitable electrical shock is presented to the nerve supplying the chosen muscle. Normally, an EPP would cause the postsynaptic, or muscle membrane, to reach threshold for an action potential which would subsequently cause the muscle cell to contract, breaking the electrode. To prevent muscle contraction, paralysis is usually maintained by lowering Ca and raising Mg levels to reduce the amount of ACh released for each presynaptic action potential. In such experiments spontaneous or evoked synaptic potentials are monitored during a suitable control period with constant superfusion of the muscle. The toxin is applied by switching the perfusing medium to one containing the appropriate concentration of the toxin and recording proceeds continuously until a suitable end point is reached or the microelectrode is dislodged from the cell.

RESULTS

Effects of Crude Endotoxins on Neuromuscular Transmission

Our initial studies employed a number of crude endotoxins prepared using the Boivin procedure by Difco Laboratories from the *Escherichia coli* serotypes 055:B5 and 0127:B8. The results of these experiments demonstrated that certain preparations elicited a biphasic pattern of alterations in MEPP frequency. A preliminary increase in frequency was followed in most cases by a profound depression of spontaneous transmitter release [17]. Figure 1 shows results from three sets of experiments performed under indentical conditions with ETX concentrations at 10 μg/ml of bath. There is a significant difference between the two samples of B8 material used, which supposedly differed only by being obtained from different production lots of the toxin. The overall effect of the active B8 and the B5 toxins was one which suggested an early facilitation of transmitter release followed by a late depression. The time course of the biphasic pattern is dose dependent and the data show that the peak frequency reached is independent of concentration. For example, with the B5 toxin, a peak frequency of 11–13 times control is reached after 20, 180, and 260 min of exposure to 10, 1, and 0.1 μg/ml, respectively. Experiments using the "active" B8 ETX were subsequently performed in media in which Ca was omitted and sufficient EGTA was added to lower the external Ca concentration to 10^{-8} M. The results of these experiments indicate that the early increase in MEPP frequency is dependent on the level of extracellular Ca and that the final "depressed" frequency observed after prolonged exposure is equivalent to that observed in Ca-free medium. Significantly, despite a drastic reduction in MEPP frequency, there is no significant change in MEPP amplitude. This indicates the absence of a significant postsynaptic effect of ETX on ACh receptor sensitivity. On the basis of these experiments, we hypothesized that ETX may act as a partial Ca ionophore serving to facilitate Ca entry during the early stage of exposure, but subsequently blocking Ca access to the membrane or inhibiting transmembrane movement of Ca with a consequent blockade of transmitter release.

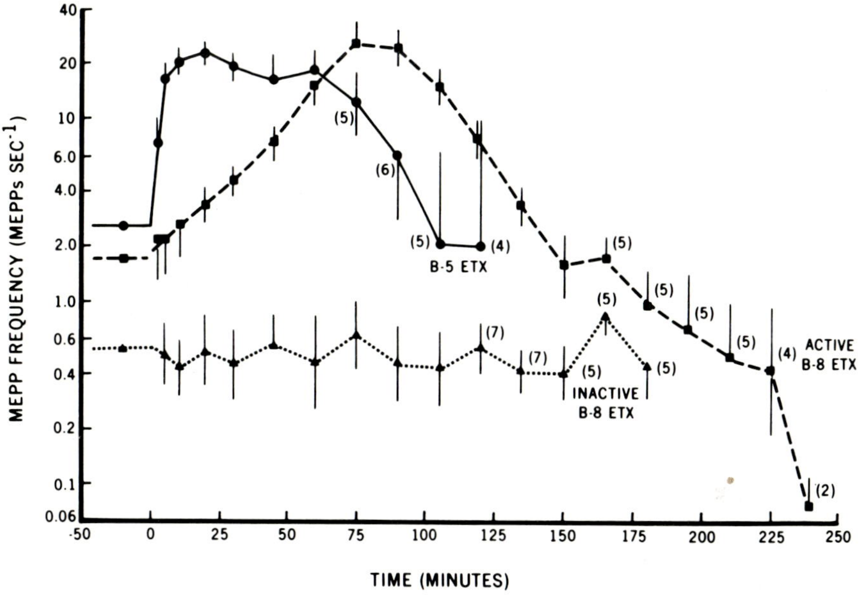

Fig. 1. Effects of 10 μg/ml of various Difco *E. coli* ETX in MEPP frequency at frog neuromuscular junctions. Continuous exposure to B5 (filled circles, N = 11), active B8 (filled squares, N = 6), or inactive B8 began at Time 0 preceded by a 45-min control period in normal Ringer's. Points represent means ± SE; numbers in parentheses note loss of N as individual experiments were terminated by loss of cell impalement ([17] reprinted by permission).

Effects of Heat-Killed *Salmonella typhimurium* on Neuromuscular Transmission

The variable effects of crude endotoxins on transmitter release as a function of ETX serotype and production lot number suggested that the portions of the macromolecule responsible for modulating transmitter release might be quite labile. Consequently, we performed experiments employing heat-killed, lyophilized *S. typhimurium* (SR-11) cultured and prepared by Dr. Rod McCallum [18]. Representative results of these experiments are depicted in Figure 2. Neuromuscular junctions were exposed for prolonged periods to 1, 10, and 100 μg/ml (only 10 and 100 μg/ml are shown). Two effects are notable: (1) At sufficiently high concentrations, unextracted *Salmonella* bacteria produce a biphasic pattern of spontaneous transmitter release similar to that seen with some extracted toxins, and (2) after a sufficient period of exposure the junction becomes impervious to the effects of the cation ionophore X537A. Exposure of junctions to this concentration of X537A (100 μM) normally provokes a catastrophic reaction in the presynaptic terminal signaled by a depolarization of the end plate area of the

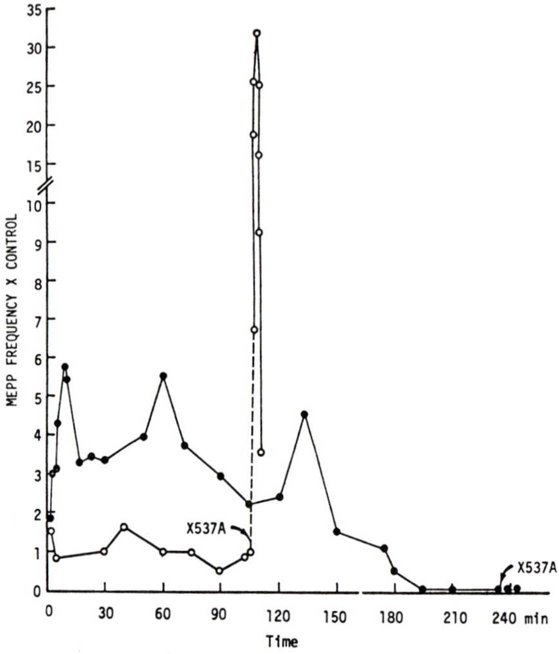

Fig. 2. Effects of heat-killed *S. typhimurium* at concentrations of 10 (open circles) and 100 μg/ml (filled circles) on MEPP frequency in representative experiments. Toxin was introduced at Time 0; subsequently the ionophore X537A (100 μM) was added at arrows. ([18]reprinted by permission).

muscle cell and a massive outpouring of transmitter similar to that shown in Figure 2 with *Salmonella* bacteria at 10 μg/ml. These results suggest that both components of ETX action, that of early acceleration of transmitter release and that of late isolation of the neural membrane can be elicited by the essentially unaltered *Salmonella* bacterial membranes. Presumably then, the particular extraction process producing crude ETX and any subsequent purification of the toxin has important consequences on the membrane toxicity of the material employed in a particular experiment.

Effects of RNA-Free Lipopolysaccharide

Changes in both spontaneous and evoked transmitter release were observed in the presence of protein-free lipopolysaccharide (LPS) derived by McCallum from *S. typimurium* (SR-11). Surprisingly, the effects on MEPP frequency were monophasic; no early facilitation of spontaneous transmitter release was observed at normal Ca concentrations [19]. However, a biphasic effect could be produced if Ca concentration was initially elevated to 10 mM (from normal at 1.8 mM). This suggests that the Ca ionophore component of the macromolecule may be partially hidden or inactivated by the polysaccharide chain in this purified form.

The usual effects on evoked release are summarized in Figure 3. Exposure to LPS at 10 μg/ml induces an abrupt decline in EPP amplitude, which is entirely a consequence of decreased quantal content of the EPPs. That is, the amount of transmitter released for each EPP is reduced while there is no parallel change in muscle cell membrane resting potential or in the amplitude of MEPPs. If the concentration of LPS is sufficiently high, evoked release is suppressed entirely, while MEPP frequency is again reduced to levels normally observed only in Ca-free medium. For a period of time after initial exposure to LPS, transmission efficiency can be restored by raising the Ca concentration of the bath. This period of reversibility roughly corresponds to the period during which X537A effects can be induced. Figure 4 shows results in four experiments in which junctions were exposed to LPS at 10 μg/ml and then at various later times the bathing media was changed to one also containing 100 μM ionophore. Up to 40–50 min of exposure to LPS caused no change in the response to the ionophore. For example, in the cell exposed to X537A after 32 min in LPS (filled triangles) MEPP frequency increased nearly 1000-fold and the cell impalement was lost during violent fasiculations of the muscle. Subsequently, the presynaptic terminal apparently becomes isolated or impervious to the ionophore; no increase in MEPP frequency is seen despite recording the usual depolarization of the muscle cell endplate area.

Results essentially similar to these are seen if the toxin used is extracted with phenol, rather than TCA, via the Westphal technique from either *E. coli* or *S. typhimurium*. The only difference between these phenol-extracted materials and the LPS resulting from further purification to remove protein is the slope of the MEPP frequency versus exposure time plot: MEPP frequency declines to the same asymptote at a faster rate in the presence of LPS.

The results seemed to indicate that there are two separable components in the action of ETX on excitation–secretion coupling. The first, and earlier component was one which facilitated transmitter secretion, possibly by facilitating Ca movement across the membrane. It is, however, labile, and did not seem to survive phenolic extraction. The second component, that which depresses transmitter release and promotes isolation of the presynaptic terminal, was present with any form of ETX.

Effects of Glycolipid from Re 595, *Salmonella minnesota*

Work now in progress employs a phenol-extracted glycolipid from the Re 595 mutant of *S. minnesota*, which produces only lipid A and the ketodeoxyoctonate (KDO) bridge to which the polysaccharide chain normally attaches in the un-

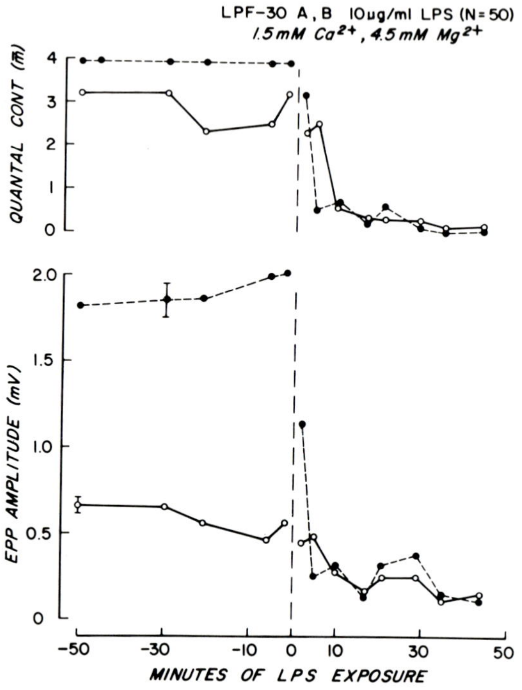

Fig. 3. Effects of 100 μg/ml of *S. typhimurium* LPS on evoked acetylcholine release at two neuromuscular junctions in the same muscle monitored simultaneously. Lower portion: EPP amplitude in muscle-paralyzing, low-calcium (1.5 mM Ca; 4.5 mM Mg) Ringer's. Each point is the mean of 50 repetitions at 0.5 Hz; largest SE for EPP at each junction is shown. Upper portion: mean quantal content (m) of EPPs calculated by method of failures [16] at the same junctions. LPS exposure began at Time 0 ([19] reprinted by permission).

mutated bacteria. Contrary to expectations, exposure of neuromuscular junctions to the glycolipid (LPA) produces a biphasic modulation of both spontaneous and evoked release: an acceleration of spontaneous release and facilitation of action potential-induced evoked release followed by typical deceleration of MEPP frequency to levels seen in Ca-free medium and blockage of EPPs. Figure 5 shows several individual experiments observing MEPP frequency over a range of 10 μg to 1 ng of LPA/ml of bath. Two effects are apparent: (1) transmitter release is either increased or depressed slowly as a function of dose down to a concentration of 10 ng/ml, and (2) at higher doses (1 and 10 μg/ml) there is a second

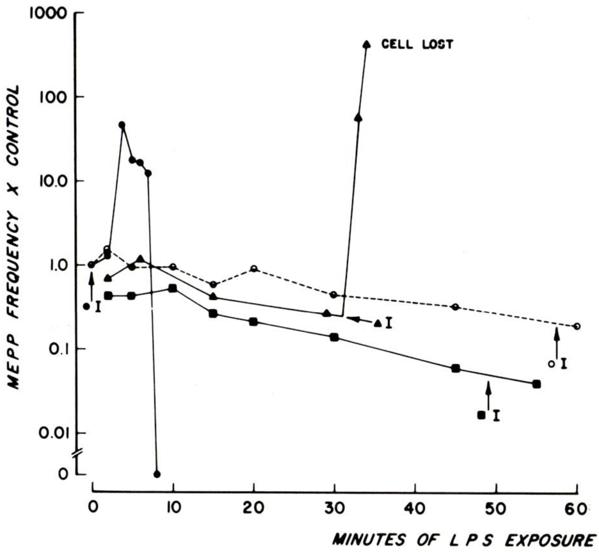

Fig. 4. Effects of ionophore X537A (100 μM) on MEPP frequency at neuromuscular junctions exposed to 10 μg/ml of LPS. Introduction of ionophore was made at times indicated by arrow and symbol I in four separate experiments ([19] reprinted by permission).

component apparent which is marked by abrupt decrease (faster decline) in the slope of MEPP rate depression with a final stabilization. With decreasing concentrations the late phase disappears.

Figure 6 shows the effect of LPA on evoked transmitter release (measured by EPP amplitude) over a concentration range of 10 μg to 1 ng of LPA/ml of bath. Again two components of action are immediately apparent: (1) a slow component of either facilitation or slight depression of evoked transmitter release and (2) a component of depression with a much more negative (faster decline) slope. At the highest concentration (10 μg/ml), EPP amplitude is elevated, indicating a possible ionophoric effect of LPA; at 80 min of exposure, there is an abrupt decline in amplitude which continues to complete blockade within a few minutes. With lower concentrations the first component is converted to a shallower slope and the appearance of a final asymptote. These data suggest that the initial, presumed Ca ionophore component of LPA action serves to keep transmitter secretion elevated; at some subsequent time the second component predominates and the membrane becomes more or less isolated from extrenal Ca. The kinetics of experiments at 1 and 10 μg/ml indicate that there can be a

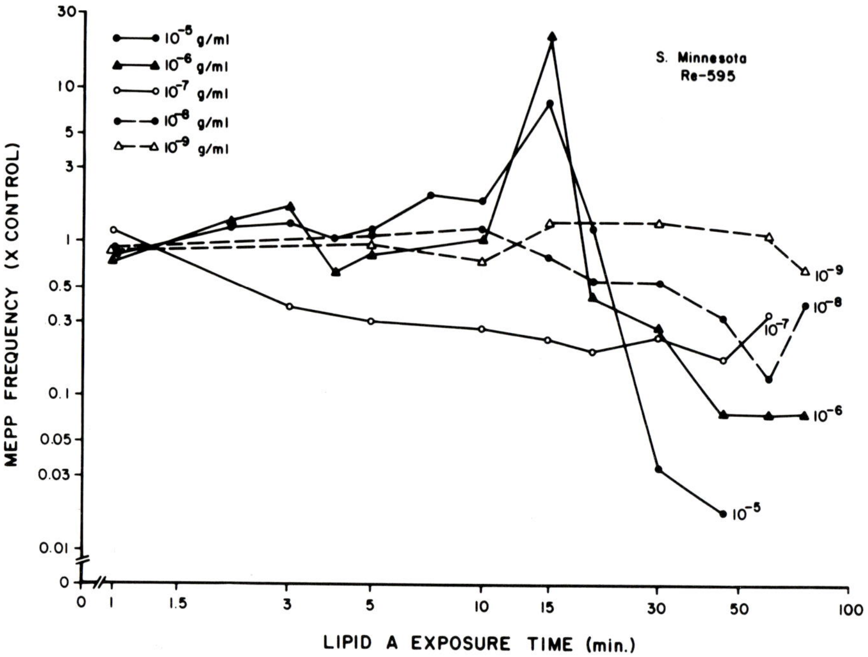

Fig. 5. Alterations in MEPP frequency at neuromuscular junctions exposed to various concentrations of *S. minnesota* Re 595 glycolipid (LPA). Data are from five experiments, one at each concentration. Note logarithmic scale on both axes of graph.

sudden transition from the first component to the second within a few minutes. In some experiments the transition may be so swift that the first component is either absent or it cannot be resolved at particularly sensitive junctions.

Figure 7 describes a single experiment in which LPA and Ca are shown to be simple antagonists. The EPP amplitude at this junction was depressed immediately by 1 μg/ml of LPA. The depression was not reversed by returning to an LPA-free control bath, but EPPs were immediately reversed to control amplitude with the addition of 0.5 mM Ca. With reintroduction of 1 μg of LPA/ml of bath in the presence of elevated Ca (now 2.3 mM), transmitter release is facilitated above control. Subsequently, when LPA concentration is raised to 10 μg/ml of bath, depression occurs immediately and EPP amplitude stabilizes and can then be reversed again by a large elevation in bath Ca (to 16.6 mM in this experiment). Such experiments suggest a clear antagonism between Ca and LPA during the early component of LPA action.

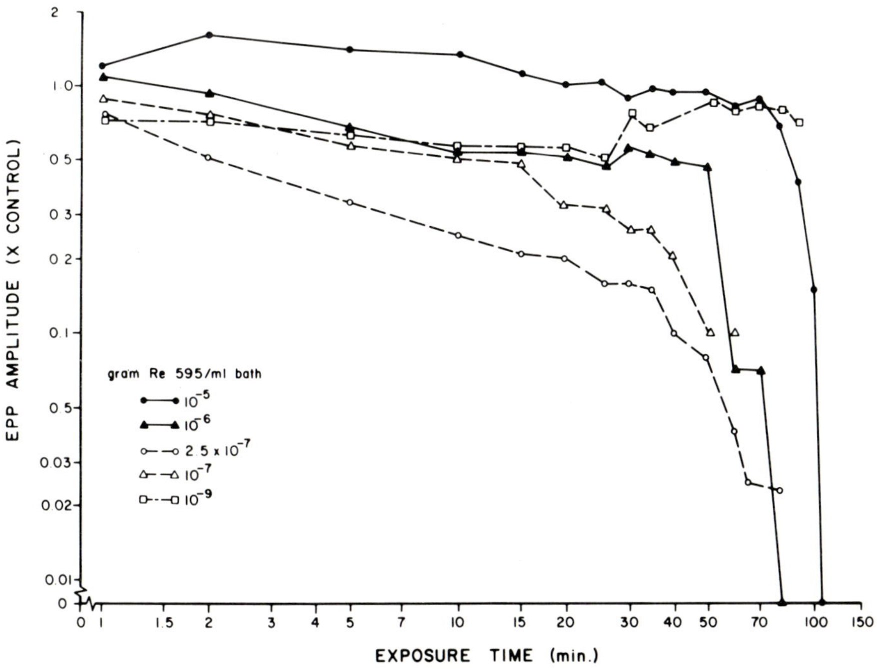

Fig. 6. Alterations in EPP amplitude at neuromuscular junctions exposed to various concentrations of Re 595 glycolipid (LPA). Data are from six experiments, one at each concentration; data points are means of 25 (1- and 2-min samples), 50 (5-min sample), or 200 EPPs evoked at 0.5 Hz. Note logarithmic scale on both axes of graph.

Measurement of quantal content (right-hand ordinate in Fig. 7) indicates that the reduction of EPP amplitude is presynaptic in origin. There are, however, suggestions in this (eg, at 15–25 min) and in other experiments that LPA, in contrast to LPS, may have a significant postsynaptic action since quantal content increases without a parallel increase in EPP amplitude.

Results of this work suggest that the lipid A moiety, at least when presented to the membrane as Re 595 glycolipid, unexpectedly produces, once again, a biphasic pattern of transmitter release, a pattern which includes the previously documented acceleration of spontaneous transmitter release, but now also includes a significant, transient elevation in evoked transmitter release. Such a significant difference between the actions of LPA and LPS is not without precedent since Morrison and Kline [20] have shown that the lipid A region of the LPS macromolecule is responsible for classical pathway activation of comple-

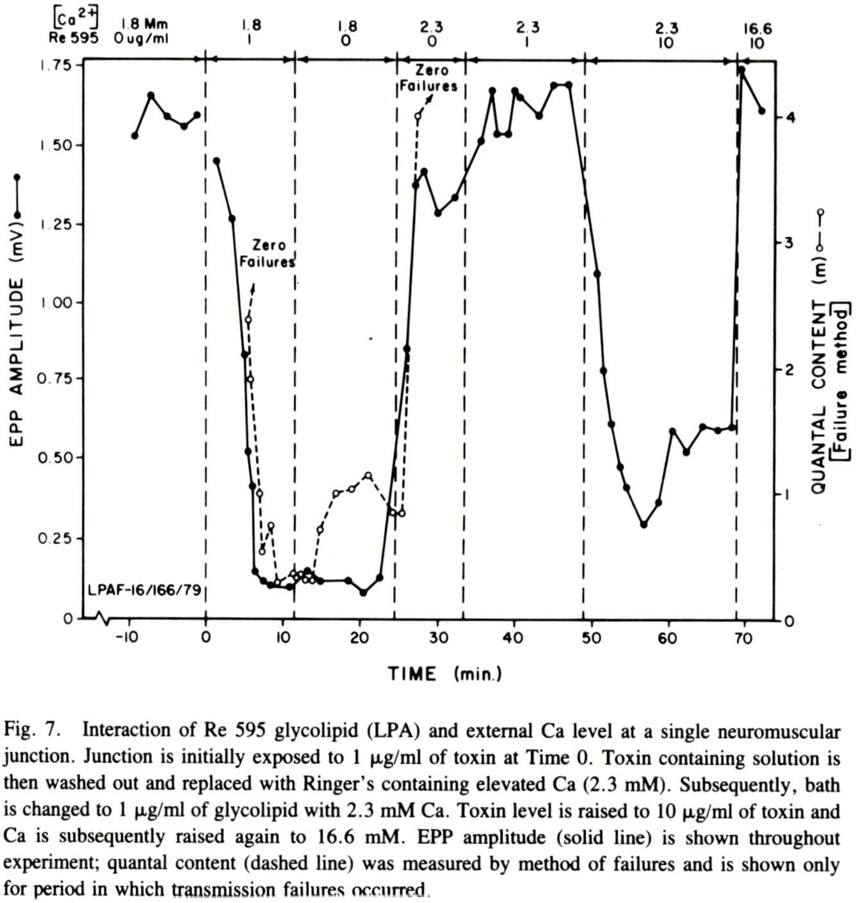

Fig. 7. Interaction of Re 595 glycolipid (LPA) and external Ca level at a single neuromuscular junction. Junction is initially exposed to 1 μg/ml of toxin at Time 0. Toxin containing solution is then washed out and replaced with Ringer's containing elevated Ca (2.3 mM). Subsequently, bath is changed to 1 μg/ml of glycolipid with 2.3 mM Ca. Toxin level is raised to 10 μg/ml of toxin and Ca is subsequently raised again to 16.6 mM. EPP amplitude (solid line) is shown throughout experiment; quantal content (dashed line) was measured by method of failures and is shown only for period in which transmission failures occurred.

ment without any dependence on antibody to lipid A, while the polysaccharide portion is responsible for properdin pathway activation independent of the lipid A portion. Furthermore, these investigators found evidence that the presence and the biochemical character of the polysaccharide portion may modulate or mask some lipid A actions [20].

DISCUSSION

We suggest that a working hypothesis of ETX action on the neural membane can be proposed: The interaction with the membrane consists of at least two stages or components. The first component (component I) involves an increase

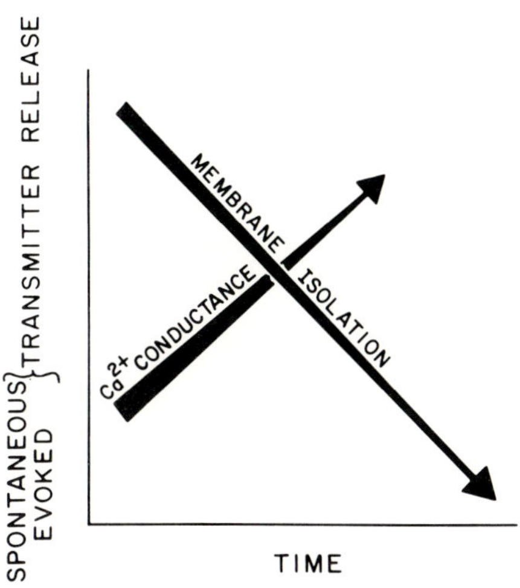

Fig. 8. Schematic diagram of two components of ETX action.

in Ca permeability across the membrane. This component is transient, subject to considerable variability in potency as a function of level of purification, purification procedure used (eg, TCA vs phenol), and contamination with miscellaneous bacterial constituents. The second component (component II) involves an inevitable and possibly irreversible isolation of the membrane from the extracellular space.

Component I is expressed at the neuromuscular junction as (1) an increase in the rate of spontaneous transmitter release which apparently derives from an increase in resting Ca permeability and (2) as an increase in the total "activated" Ca conductance induced by membrane depolarization, which results in a facilitation of evoked transmitter release. In other systems, component I should be expressed as increased levels of spontaneous or evoked hormone release, increased tone in smooth muscle, and increased contractility in both smooth and cardiac muscle as a function of their respective almost complete or partial dependence on extracellular Ca, and increased autonomic drive due to facilitation of ganglionic and end organ neurotransmission.

Component II is expressed in this experimental model as a decreased rate of spontaneous transmitter release, as a depression or blockade of evoked transmitter release, by the inability of potent cation ionophores to provoke increased transmitter release or restore it after total blockade, and the inability of increased extracellular Ca to increase or restore transmitter release. Component II should be expressed in other systems as decreased hormone secretion to previously adequate physiological stimuli, decreased tone in spontaneously active smooth muscle, decreased contractility in smooth and cardiac muscle (again as a function extracellular Ca requirements), a depression of autonomic reflexes, and generally as a debilitation of all peripheral Ca-dependent processes. However, the specificity of component II for Ca-dependent processes is as yet untested.

A schematic portrayal of the dual component mode of ETX action is shown in Figure 8. It should be emphasized that the specification of two components in the action of ETX may well be a convenient fiction, but it serves as a good working hypothesis. It is conceivable that the two components are expressions of a continuous process of rearrangement of the neural membrane during a prolonged interaction with ETX, the eventual end point of which is membrane isolation. Some of the data we have developed suggest that LPA may attach at or near the membrane site of the normally present "slow Ca channel" ordinarily activated by membrane depolarization. The act of LPA attachment could first distort the channel, transiently increasing Ca permeability, while at the same time a process of membrane transformation, substitution, or isolation is initiated which eventually reduces Ca permeability to nil.

Our work on ETX components and excitation–secretion coupling fits well with the few previous studies which have examined indirectly the effects of systemic and local administration of toxins on excitable membranes [3–10] but it is at some variance with the results obtained in direct exposure experiments by others [11, 12]. Such discrepancies as do exist are likely to be a result of the specific ETX preparation used. But such discrepancies emphasize the gaps in our understanding of the basic membrane events associated with the production of shock. We have described significant neurotoxic components in the ETX complex and we have suggested how such neurotoxicity may be expressed clinically in the initiation of septic shock and its progression to an irreversible condition.

ACKNOWLEDGMENTS

This work was supported by the Biophysics Program, Office of Naval Research, Department of the Navy Contract N00014-77-C-0630, Task NR 202-160.

REFERENCES

1. Hinshaw LB: Release of vasoactive agents and the vascular effects of endotoxin. In Kadis S, Weinbaum, G (eds): "Microbial Toxins." New York: Academic Press, 1971, p 209.
2. Hinshaw LB, Vick JA, Jordan MM, Wittmers LE: Vascular changes associated with development of irreversible endotoxin shock. Am J Physiol 202:103, 1962.
3. Anderson FL, Jubiz W, Tsagaris T, Kuida H: Endotoxin-induced prostaglandin E and F release in dogs. Am J Physiol 228:410, 1975.
4. Rao PS, Bhagat BD, Cavanagh D: Effect of endotoxin on hemodynamics and norepinephrine metabolism. Proc Soc Exp Biol Med 141:412, 1972.

5. Fine J: Some observations on the mechanisms of action of bacterial endotoxins. In Landy W, Braun W (eds): "Bacterial Endotoxins." New Brunswick: Rutgers University Press, 1964, p 585.
6. Pohorecky LA, Wurtman RJ, Team D, Fine J: Effects of endotoxin on monamine metabolism in the rat. Proc Soc Exp Biol Med 14:739, 1972.
7. Bitensky MW, Gorman RE, Thomas L: Selective stimulation of epinephrine-sensitive adenyl cyclase in mice by endotoxin. Proc Soc Exp Biol Med 138:773, 1971.
8. Palmerio C, Zetterstrom B, Shammash J, Euchbaum E,Frank E, Fine J: Denervation of the abdominal viscera for the treatment of shock, N Eng J Med 269:709, 1962.
9. Reilly J, Rivalier E, Compagnon A, Laplane R, du Buit H: Sur la pathogenie de la dothienenterie. La fievre typhoide experimentale. Ann Med 37:182, 1935.
10. Rielly J, Rivalier E, Compagnon A, Laplane R, du Buit H: Sur la pathogenie de la dothienenenterie. Le role du systeme neurovegatatif dans les genese des lesions intestinales. Ann Med 37:321, 1935.
11. Parnas I, Reinhold R, Fine J: Synaptic transmission in the crayfish: Increased release of transmitter substance by bacterial endotoxin. Science 171:1153, 1971.
12. Conner J, Fine J, Kusano K, McCrea MJ, Parnas I, Prosser CL: Potentiation by endotoxin of responses associated with increases in calcium conductance. Proc Nat Acad Sci USA 70:3301, 1977.
13. Hess ML, Briggs FN: The effect of Gram-negative endotoxin on the calcium uptake activity of sarcoplasmic reticulum isolated from canine myocardium. Bioc Biophys Res Commun 45:917, 1971.
14. MacNicol MF, Goldberg AH, Clowes GHA: Depression of isolated heart muscle by bacterial endotoxin. J Trauma 13:554, 1974.
15. Rubin RP: "Calcium and the Secretory Process." New York; Plenum Press, 1974.
16. Hubbard JI, Llinas R, and Quastel DMJ: "Electrophysiological analysis of synaptic transmission." Baltimore: Williams and Wilkins, 1969.
17. Person RJ: Endotoxin alters spontaneous transmitter release at the frog neuromuscular junction. J Neurosci Res 3:63, 1977.
18. Person RJ: Depression of spontaneous and ionophore-induced transmitter release by *Salmonella*. Experientia 35:801, 1979.
19. Person RJ: Bacterial lipopolysaccharide depresses spontaneous, evoked, and ionophore-induced transmitter release at the neuromuscular junction. J Neurosci Res 4:105, 1979.
20. Morrison DC, Kline LF: Activation of the classical and properdin pathways of complement by bacterial lipopolysaccharides. J Immunol 118:362, 1977.

Some Effects of *Escherichia coli* Endotoxin on Cells in Culture

Ian A. Silver

Although the actions of endotoxins on the whole animal [1–3] and clinical patients [4,5] are relatively well characterized, less is understood about the immediate effect of these complexes at the cellular level [6]. Ultrastructural changes have been recorded in cells of various organs such as the liver [7,8] in animals which have died from endotoxic shock, and it is also a common clinical finding that there is a rise in plasma potassium following the hyperglycemic phase of endotoxin, when this is succeeded by a lactate acidosis and hypoglycemia.

Because of the complex interactions of the protective mechanisms which occur during septic shock in the whole animal, it was decided to examine the direct effects of endotoxins on cells in culture. A number of different types of cells were used including BHK fibroblasts, L cells, primary cultures of fibroblasts from fetal horse lung, and neuroblasts from fetal mouse brain, together with the mouse neuroblastoma line C-1300. The purpose of the investigation was to discover how quickly endotoxin attached to the cells, how rapidly the attachment became irreversible, and whether such attachment could be delayed or prevented. Once attachment had taken place, it was hoped to find whether further protection could be provided either by the use of antiendotoxin globulin or by steroids.

MATERIALS AND METHODS

Cell Cultures

Most types of cells were grown in RPMI 1640 medium supplemented as necessary with 10% fetal calf serum and pencillin and streptomycin. The cultures were grown under standard conditions in humidified 5% CO_2 in air at 37°C. For maintenance and growth purposes, cells were placed in Falcon flasks, but when they were required for acute experiments, they were detached mechanically from the flask and suspended in culture medium. They were then seeded into petri

Pathophysiological Effects of Endotoxins at the Cellular Level, pages 81 – 95

dishes until they had attached. The fibroblasts and fibroblastic lines grew best in Glasgow-modified Eagles medium with 10% tryptose phosphate broth and 10% fetal calf serum.

Endotoxin

The endotoxin used was either a commercial lipopolysaccharide from *Escherichia coli* 0111:B4 or 0127:B8 (Difco) extracted by the Westphal procedure or a lipopolysaccharide obtained by phenolic extraction from a local culture of *E. coli* 0111:B4. Test incubations of the cells were carried out at concentrations of endotoxin between 0.001 mg/ml and 1.0 mg/ml and a median lethal dose (LD_{50}) was determined in a standard culture of BHK fibroblasts for each batch of endotoxin. Some cells were incubated with and without endotoxin in the presence of a soluble steroid, dexamethasone phosphate, at steroid concentrations from 0.001 mg/ml to 1.0 mg/ml. Control experiments were carried out by incubating cells without the endotoxin.

Measurements of Ions

The concentration of potassium at the surface of the cells was measured with ion-selective microelectrodes containing a liquid ion-exchange resin (Corning 477317). Construction of the electrodes was as described previously by Zeuthen et al [9]. A double-barrel configuration was used in which the reference electrode formed part of the probe. Intracellular and extracellular measurements of potassium activity were made with these electrodes. Hydrogen ion concentration was measured similarly but with single-barreled microelectrodes of the type described by Thomas [10]. These microelectrodes had a very fine resolution and could be used to measure not only the pH at the surface of an individual cell but could also detect pH differences between various zones on the surface of a single cell which was spread out on the floor of the culture dish.

Potassium electrodes were calibrated in 1.5, 15.0, and 150.0 mM KCl and pH probes in buffers from pH 5.0 to 8.0. The reference electrode was Ag–AgCl in 0.5 mM KCl connected by an agar bridge of 4% agar in culture medium to the petri dish.

Labeled Antibody

Antiendotoxin serum was raised in guinea pigs by injection of the appropriate antigens in Freund's complete adjuvant. The *E. coli* antisera were absorbed with *Salmonella* endotoxin. The gammaglobulins were extracted from the serum by standard methods on a Sephadex column and the antiendotoxin globulin was labeled with fluorescein isothiocynate (FITC). The labeled globulin was tested for specificity and controls included studies in which normal cells were exposed to the FITC-labeled antiendotoxin and the use of globulin from nonimmunized guinea pigs. After washing, the cells were examined microscopically under ultraviolet light and showed virtually no background fluorescence.

Experimental Procedures

Investigations were carried out by exposing cells of different types to the various concentrations of endotoxin (LD_{50} X 0.1, 0.5, 1.0, 2.0, and 5.0). Subsequently, those same concentrations of endotoxin were repeated with new cultures, but before or after the addition of dexamethasone phosphate into the medium. In addition, other experiments were performed in which appropriate quantities of antiendotoxin globulin were added to the cultures before the endotoxin and in a final study, the antiendotoxin globulin was added to the culture at various periods after the endotoxin had been given. The cells were incubated after the addition of the higher concentrations of endotoxin for 5, 10, 15, 30, 60, 90, 120, and 240 min and after low concentrations for 2, 12, 18, and 24 hr. Cell samples were taken at each of these stages and prepared for electron microscopy (see below). At the same time, measurements were made of intra- and extracellular potassium activity and the rate of K^+ loss from the surface of the cells. In a few cases, membrane potentials were determined by measurement of the DC voltage on the reference barrel of the double-barreled potassium electrode.

Microscopy Procedures

For electron microscopy cells were fixed in cold buffered (pH 7.4) glutaraldehyde and stored at 4°C. They were washed with cacodylate buffer, and postfixed in osmium, dehydrated, and then cleared in propylene oxide. The cells were embedded in epoxy resin (Epon 812) and sections were cut at 80–100 nm. These sections were mounted on grids and examined in a Phillips EM201 electron microscope.

Those cells which had been exposed to antiendotoxin globulin were washed twice in fresh culture medium and examined under epi- and transillumination with a Leitz ultraviolet microscope. Cells were divided into two batches: those which were treated with antiendotoxin globulin before fixation and those which were so treated after fixation with cold acetone. Antibody was applied to the cells after fixation because in this way the plasma membrane could be rendered permeable to large molecules and the presence of endotoxin in the deeper parts of the cell might be visualized if it were still antigenic.

Whole Animal Studies

The endotoxin batches were also tested in rats. The lipopolysaccharide was dissolved in physiological saline and administered intraperitoneally to anesthetised white Wistar-derived rats at a rate of 2 mg/100 gm. The blood pressure was monitored, and when it fell terminally (usually 6–8 hr after endotoxin

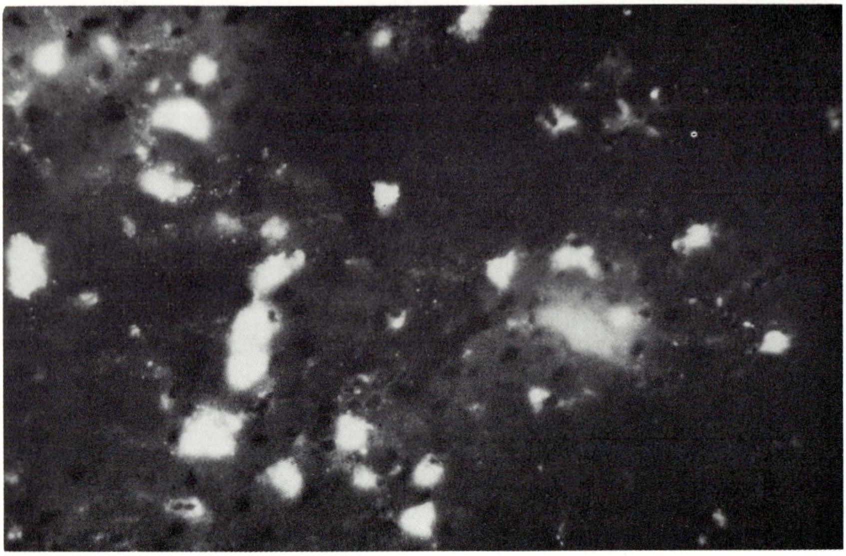

Fig. 1. Frozen section of liver from a rat given 2 mg/100 g of body weight endotoxin, stained with FITC-labeled antiendotoxin globulin. Note fluorescence in Kupffer cells, and leakage into parenchyma (× 450).

administration) tissues were taken for histology into neutral buffered formalin (10%), and for electron microscopy into 4% ice-cold glutaraldehyde. Frozen sections were prepared of the liver and flooded with antiendotoxin antibody to demonstrate uptake of endotoxin by Kupffer cells (Fig. 1). The carotids of some rats were injected with a casting resin, methyl methacrylate. Subsequently, the tissue was macerated and the casts were shadowed with gold and photographed under a scanning electron microscope in the manner described in reference [13].

RESULTS

The general findings were that the initial effects of endotoxin varied markedly from one type of cell to another. Cells derived from primary cultures (the equine lung fibroblasts and the mouse neuroblasts) were particularly susceptible to the endotoxin, whereas the transformed cells grown as commercial lines were much more resistant.

Morphological changes appeared in the majority of cells only after between 60 and 90 min of exposure to lethal concentrations of endotoxin, although the cells were damaged more quickly if exposed to high concentrations of steroids.

As is customary with endotoxin, the concentration of toxic material required to produce a particular level of cell damage varied from batch to batch. The most toxic batch had a biological activity which was approximately 80 times higher

TABLE I. Changes in Some Cultured Cells Caused by Administration of *E. coli* Endotoxin at 2 LD_{50} (24 hr)

Cell type	Normal membrane Potential (−mV ± SEM)	Time after 2 LD_{50} (24 hr) endotoxin administration for change of membrane potential	Recovery if endotoxin washed off at 15 min	Potassium leakage detectable after endotoxin as in column 3
Rat fibroblasts	58 ± 6	1 hr	+ +	40 min
Mouse L cell line	20 ± 5	20 min	+ + + +	15 min
C1300, NAB 41 mouse neuroblastoma line	78 ± 7	3 hr	+	3 hr
Macrophages (rabbit)	18 ± 8	No change	+ + + +	No leakage
Primary fetal mouse neuroblast	75 ± 6	10 min	No recovery	10 min

than that of the least effective batch on a dry weight basis. Because of this variability, it was decided to adopt a standard endotoxin dose which gave approximately 1 LD_{50} after 24 hr, or with the higher doses, a particular degree of mitochondrial damage after a period of 1.5 hr. This concentration of a particular batch of endotoxin was then used in appropriate dilutions.

Membrane Potential

Measurement of membrane potentials with microelectrodes showed a range from about 20 mV (L cells) to 80 mV (neuroblastoma cells). These correlated with values from previous studies [11]. In the presence of low concentrations of endotoxin, the potentials tended to rise initially and fall later. So far as potassium flux was concerned, some cells (L cells) showed a remarkable propensity to lose potassium at a very early stage of exposure to endotoxin, but nevertheless, retained the ability to recover. The neuroblastoma cells, on the other hand, were very resistant to depolarisation by endotoxin and leaked potassium only at a late stage when irreversibly damaged. These data are summarized in Table I.

Morphological Findings

Endotoxins caused similar changes in cultured cells to those reported for parenchyma in intact organs [7,12]. These were cellular swelling, and in particular, hydration of the endoplasmic reticulum and of the mitochondria. The mitochondrial cristae were disrupted, nuclei became crenated in appearance, ribosomes became detached, and in the late stages of endotoxin attack, the cells collapsed and leaked their contents. Figure 2 illustrates typical findings. Most morphological changes appeared between 45 and 120 min after application of endotoxin in high concentration, and after 90 min, irreversible changes were seen in all cultures treated with 2 LD_{50} or higher levels of endotoxin.

Fluorescent Antibody Findings

Cells which had been exposed to endotoxin took up fluorescent antibody on their surfaces if they were exposed to the antibody within 2 hr of the endotoxin being applied to them (Fig. 3). However, if low but still lethal concentrations of endotoxin (0.5 LD_{50}) were provided, the complex was taken up by the cells, and after about 4 hr, antiendotoxin globulin failed to show the presence of the endotoxin on the surface of most of the live cells. Nevertheless, when these same cells were fixed and then exposed to antiendotoxin globulin, many became fluorescent (Fig. 4). This suggested that the endotoxin was at first attached to the surface of the cells and then became internalized through the membrane where the antigenic structure ceased to be recognizable by the antibody. At a late stage of intoxication, the endotoxin once again became readily visualized by the application of antiendotoxin globulin, even to the live cells. This probably

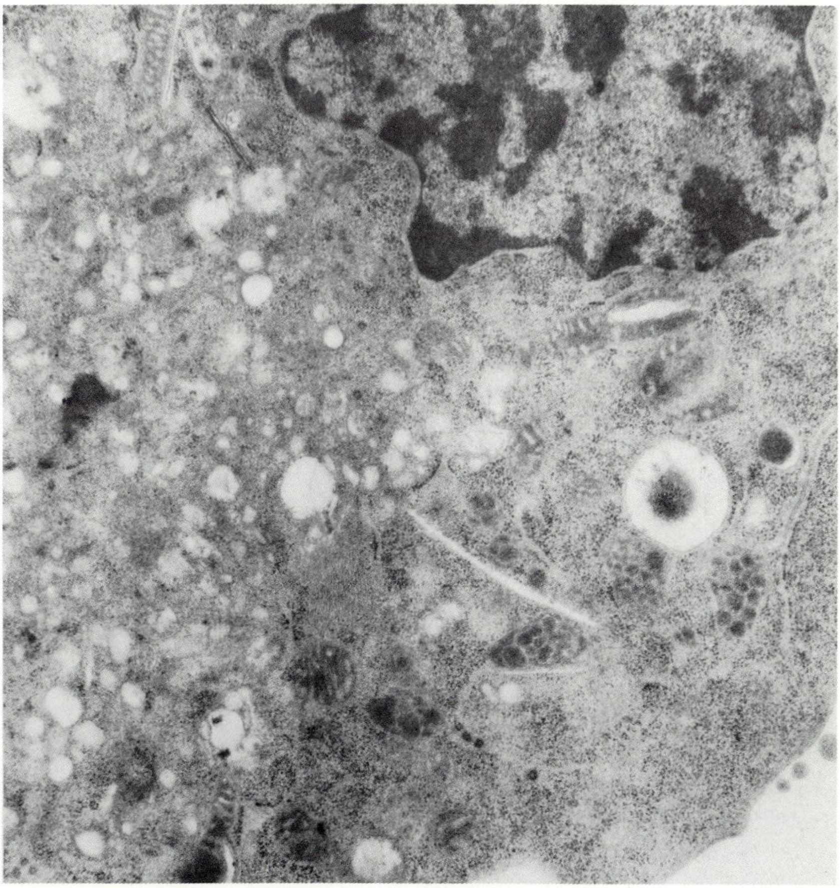

Fig. 2. TEM photograph of part of a C1300 neuroblast cell after 120-min exposure to 2 LD_{50} *E. coli* endotoxin. Note crenation of nucleus with chromatin clumping, swelling, and disruption of mitochondrial cristae, vacuolation of cytoplasm, free ribosomes, and general loss of definition of organelles (× 30,500).

represented a state where the cell membrane had become very leaky and permitted the passage even of very large molecules such as gammaglobulins, particularly since several intracellular structures became fluorescent, including lysosomes and the nuclear membrane. This apparent permeability change correlated with a fall in membrane potential and K^+ leakage.

Effects of Steroids

In cultures given steroids and lower doses (0.5 LD_{50}), the endotoxin appeared to be unable to attach to the surface of the cells, or if it did attach, was unable

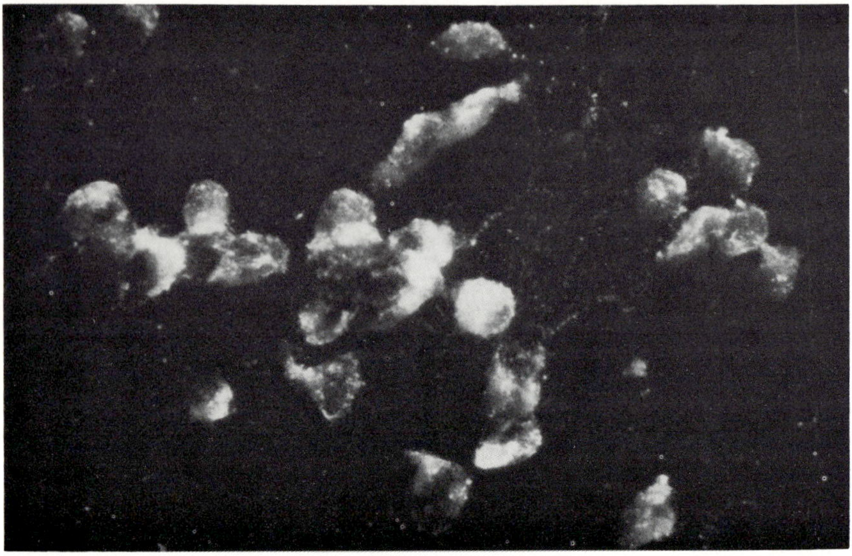

Fig. 3. Cells incubated for 2 hr with endotoxin and exposed alive to fluorescent antiendotoxin globulin (× 500).

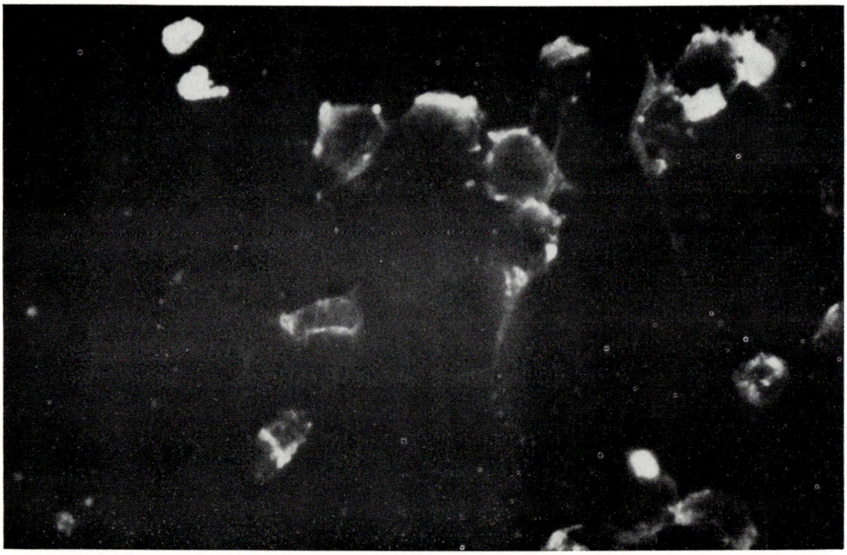

Fig. 4. Cells fixed after 18 hr following exposure of 0.1 LD_{50} endotoxin and stained with FITC-labeled antibody. When alive these cells did not take up antibody (× 500).

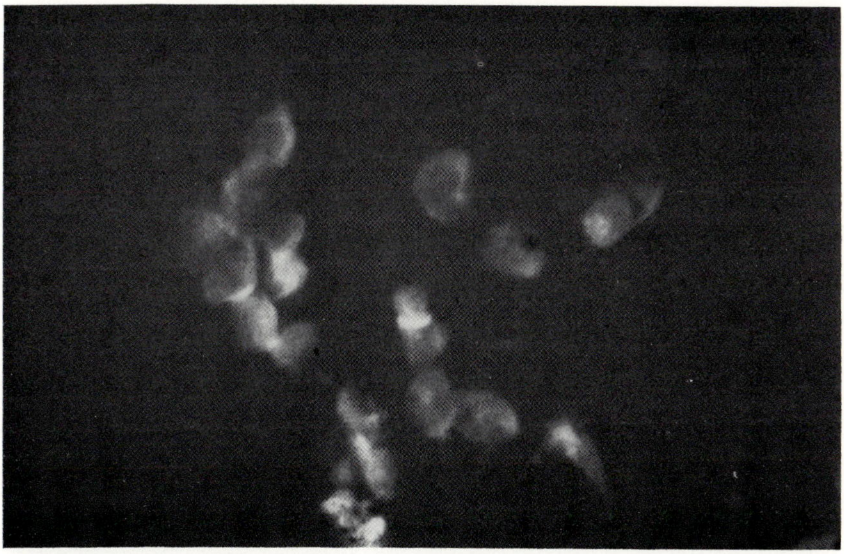

Fig. 5. Cells incubated with 0.1 mg/ml of dexamethasone phosphate before 18-hr exposure to 1 LD_{50} endotoxin and then fixed and treated with fluorescent antiendotoxin globulin. There is very little endotoxin visible (× 500).

to internalize. Figure 5 shows the appearance of cells which had been treated with 0.05 mg/ml of dexamethasone phosphate about 15 min before the administration of endotoxin (equivalent to 2 LD_{50} doses). It can be seen that there is minimal fluorescence on these cells which indicates either that the endotoxin had become internalized or that it was not attached to the cell surface. Subsequently, these cells were fixed and flooded with the fluorescent antibody, but no visible increase in antibody binding could be detected.

If the steroid was applied following endotoxin administration, it appeared to slow the structural changes that occurred with lethal doses of endotoxin, but it did not protect the cells completely. It did not cause endotoxin already bound to cells to detach (Fig. 6). Its effect on internalization of fluorescent label were inconsistent from one experimental to the other.

With regard to the effect of steroids on leakage of potassium from the surface of the cells, the findings paralleled those with the fluorescent antibody. The cells which had been given steroid before they were exposed to endotoxin did not leak potassium and, subsequently, did not show changes of mitochondrial structure. Those cells which had been exposed to the endotoxin before the steroid was administered showed approximately the same time scale of potassium leakage, as if no steroid had been given, with the exception that cells given low

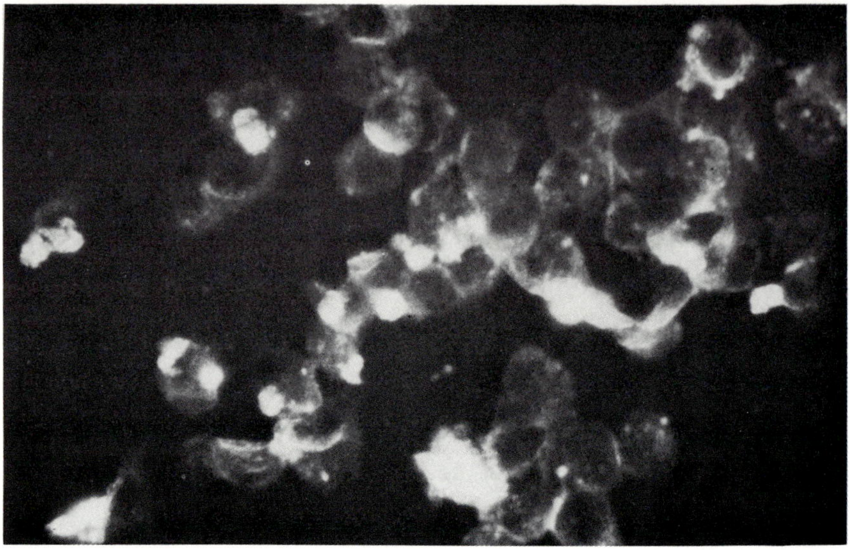

Fig. 6. Cells treated first with endotoxin and then given steroid (times and doses as in Fig. 5). Note strong fluorescence in many cells (× 500).

lethal doses of endotoxin in the presence of relatively high doses of steroid (0.05–0.1 mg/ml) appeared to be protected and subsequently showed either no potassium leakage or very minor changes from which they recovered after a few hours.

pH Changes

A decreased pH at the surface of all cells given endotoxin could be detected after about 1 hr and suggested that there had been an increase in anaerobic metabolism in the cells despite the fact that the cultures were well oxygenated.

Reversibility of Endotoxin Attachment

In some types of cells, particularly neuroblastoma and fibroblast lines, it seemed possible that the attachment of the endotoxin was less firm than with some other cell types (particularly primary neuroblast cells). In the former, if the endotoxin was administered in reasonably low concentrations (less than 2 LD_{50}), it attached to the cells within the first 10 to 15 min, but if the cells were then washed and incubated in fresh medium, they would recover and subsequently show no damage, either physiological or morphological. If such cells were exposed to antiendotoxin globulin, after a period of 0.5 hr, no endotoxin could be detected, or at the most only one or two cells showed fluorescent activity. In distinction to this, the primary neuroblasts appeared to hold the endotoxin very firmly even when they had been exposed to it for only 5 min, and washing did not remove it from the surface of the plasma membrane. Such cells inevitably passed into the dying phase even if given steroid and antiendotoxin globulin.

The Protective Effect of Antiendotoxin Globulin

Some control cells were exposed to antiendotoxin globulin alone, and no changes were observed in their morphology or physiology. The presence of excess antibody in the culture medium protected all types of cells against addition of endotoxin, although some antibody–antigen complexes appeared to be taken up by phagocytosis, particularly by fibroblastic cells.

By contrast, if cells were first exposed to endotoxin and subsequently treated with antibody, very little protection was afforded unless the antibody was given within about 2 min of the administration of endotoxin. The primary neuroblast cells were the least protected and the transformed cells the most. In the case of the C-1300 neuroblastomas and the BHK fibroblasts, excess antibody administration up to 15 min after the cells were exposed to endotoxin provided almost complete protection.

Whole Animal Response

While the toxicity of the different batches of endotoxin varied greatly in the cell cultures, they were much more uniform in the effects produced in the rats. In all cases, a dose of 2 mg/100 gm was lethal after 4–10 hr. Equally, in all cases massive localization of endotoxin in the mononuclear phagocytic system (the reticuloendothelial system) was demonstrable with fluorescent antiendotoxin globulin (Fig. 1). In some rats, endothelial changes could be seen from SEM photographs of vascular casts. (Fig 7a,b).

DISCUSSION

These experiments demonstrated that endotoxin attaches quickly to the surface of most types of cells and may be held there for periods of between 10 min and 3 or 4 hr before it is internalized through the membrane, when it becomes impossible to demonstrate its presence in the live cell with fluorescent antiendotoxin globulin. The experiments also showed that the presence of steroids had a protective effect only if the steroid was administered in the very early stages of endotoxin attack and seemed to have little useful action once the endotoxin complex had become internalized. It was also found that morphological changes occurred relatively late in endotoxin toxicity and by the time they appeared there had almost certainly been lethal changes in the metabolism of the cell. The experiments also showed that various types of cells had markedly differing susceptibility to endotoxin which was manifested not only by differences in the

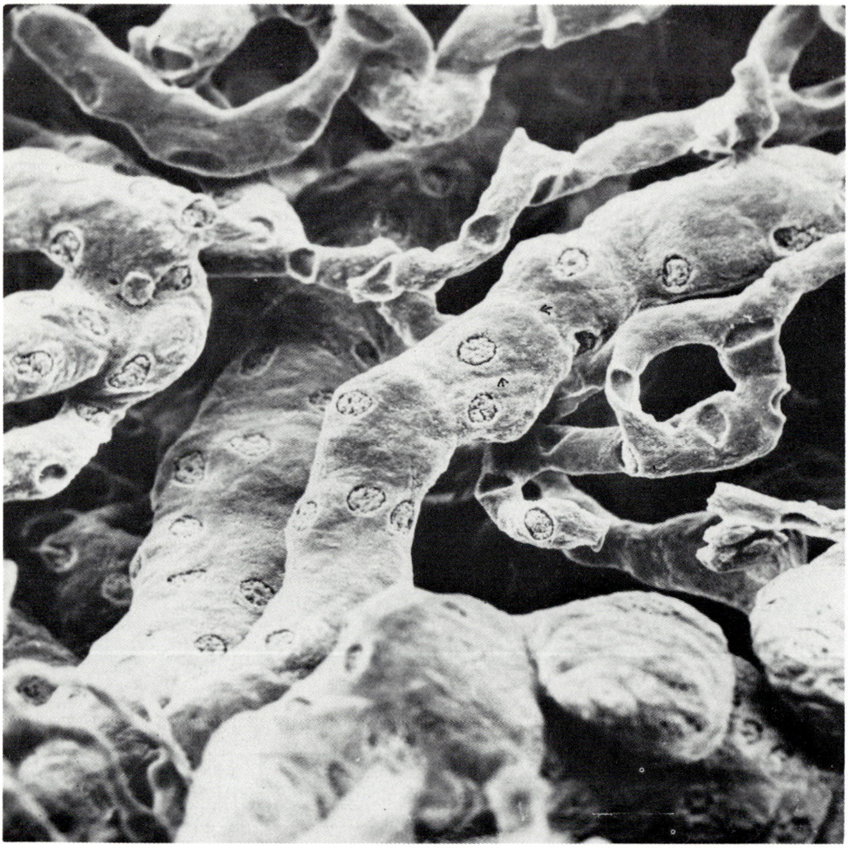

a

Fig. 7. (a), SEM of corrosion cast of normal small blood vessels in rat brain. Note well-defined endothelial cell nuclear outlines (appearing in negative relief in the cast) and endothelial cell junctions (arrowed) (× 2,000). (Courtesy of Dr. P.M. Young). (b), SEM of vascular cast from rat brain after 6 hr of endotoxemia. Note blebbing of endothelium (× 1,250).

speed with which endotoxin attached to them and was internalized, but also whether or not the attachment was firm or weak in the early stages.

The other feature that distinguished different types of cells was whether leakage of potassium occurred early or late following attack. Leakage may be produced either by the promotion of excessive permeability of the membrane by relatively large holes being made in it by some action of the endotoxin molecule, or by the endotoxin inhibiting or reducing the efficiency of the ion pumps even when the membrane remains relatively intact. It is not at all clear which mechanism, if either, is responsible for the greater leakiness of some cells as opposed

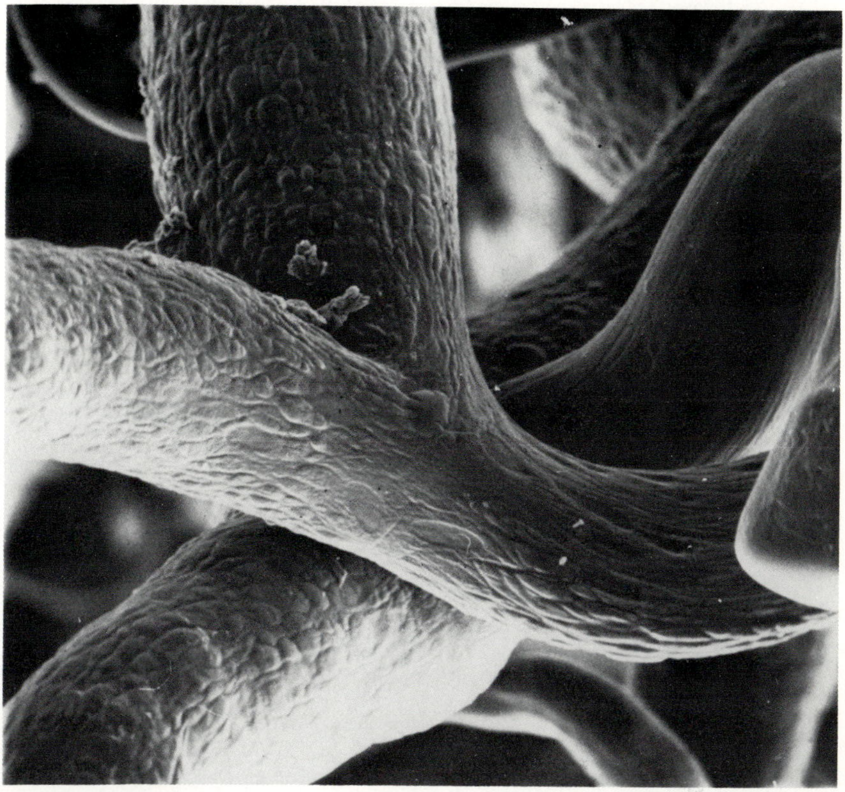

b

to others. No clear morphological evidence was obtained to support the idea of 'holes' being punched in the plasma membranes by the endotoxin.

The use of antiendotoxin globulin, although it may provide interesting insights into the antigenic portion of the endotoxin, does not necessarily give hard data on the distribution of the toxic lipid A fraction of the complex. It is quite possible that upon attachment to the cell membrane, there is a separation of the antigenic polysaccharide and the lipid A so that, although the antigen may remain attached to the membrane, the lipid A is free to diffuse into the cell cytosol. This does in fact seem likely because in recent work (Smith, Erecinska, and Silver; unpublished observation) it has been shown that metabolic changes could be found in some cells within 15 min of exposure to endotoxin. This was long before any morphological change had taken place and it was also during the period when the fluorescent antibodies appeared to show that the antigenic part of the endotoxin was still firmly bound to the plasma membrane.

What these experiments do demonstrate however, is that attachment of the endotoxin to the plasma membrane occurs within minutes and may explain why the protective effect of steroid is noticeably less if the drug is given after exposure to endotoxin than if it is given early or before endotoxemia develops [14]. The only protective effect that was observed with steroids given after endotoxin attachment to the cell had occurred was when the endotoxin was present in a relatively low dose (less than 2 LD_{50}). Thus, if one can extrapolate from a culture to a whole animal or to a human patient, steroids may be useful in protecting those cells which have either not yet been attacked or which have been attacked by only a few endotoxin molecules, whereas it affords little protection for those cells which have suffered massive invasion by endotoxins.

Finally, it was of interest that the toxicity of a particular batch of lipopolysaccharides was not equally expressed in the whole animal and the cultures. It was particularly noticeable that the less soluble batches were of relatively low toxicity in cultures, but equally, or even more, toxic to the whole animals than the more soluble batches. Clearly, processing of endotoxin in the body is an important factor in enhancing the toxicity of the complex. The relationship between particle size and biological activity of *E. coli* endotoxin has been commented on by Beer, Braude, and Brinton [15].

REFERENCES

1. Kun E, Miller CP: Effect of bacterial endotoxins on carbohydrate metabolism of rabbits. Proc Soc Exp Biol Med 67:221, 1948.
2. Wolfe RR, Burke JF: Glucose and lactate metabolism in experimental septic shock. Am J Physiol 235:219, 1978.
3. Morrison DC, Ulevitch RJ: The effect of endotoxins on host mediation systems. Am J Pathol 93:527, 1978.
4. Weil MA, Shubin H, Biddle M: Shock caused by Gram-negative microorganisms: Analysis of 169 cases. Ann Intern Med 60:384, 1964.
5. Shumer W, Das Gupta TK, Moss GS, Nyhus LM: Effect of endotoxaemia on liver cell mitochondria in man. Ann Surg 171:875, 1970.
6. Bradley SG: Cellular and molecular mechanisms of action of bacterial endotoxins. Ann Rev Microbiol 33:67, 1979.
7. De Palma PG, Harano Y, Robinson AO, Holden WD: Structure and function of hepatic mitochondria in haemorrhage and endotoxaemia. Surg Forum 21:3, 1970.
8. Blair O, Stenger R, Hopkins R, Simeone F: Hepatocellular ultrastructure in dogs with hypovolemic shock. Lab Invest 18:172, 1968.
9. Zeuthen T, Hiam RC, Silver IA: The recording of ion activities in brain with ion selective micro-electrodes. In Berman HJ, Herbert NC (eds): "Ion Selective Microelectrodes." New York: Plenum Press, 1974, p 98.
10. Thomas RC: Intracellular pH of snail neurones measured with a new pH sensitive glass microelectrode. J Physiol (London) 238:159, 1974.
11. Deutsch C, Erecinska M, Werrlein R, Silver IA: Cellular energy metabolism, transplasma and transmitochondrial membrane potentials and pH gradients in mouse neuroblastoma. Proc Nat Acad Sci 76:2175, 1979.

12. White R, Mela L, Bacalzo L, Olofsson K, Miller L: Hepatic ultrastructure in endotoxaemia, haemorrhage and hypoxia: Emphasis on mitochondrial changes. Surgery 73:523, 1973.
13. Young PM, Silver IA, Robertson-Rintoul J, Smith EM: Scanning electronmicroscopy of microvascular casts. 1. Improved penetration and resolution with controlled polymerization. J Anat (London): in press.
14. Shumer W: Steroids in the treatment of clinical septic shock. Ann Surg 184:333, 1967.
15. Beer H, Braude A, Brinton C: Relationship between particle size and biological activity of *E. coli* Boivin endotoxin. J Clin Invest 44:592, 1965.

Metabolic Effects of Endotoxins

Hepatocyte–Kupffer Cell Interactions in the Inhibition of Hepatic Gluconeogenesis by Bacterial Endotoxin

Roderick E. McCallum

Septic shock remains one of the primary problems in medicine today. Unchanged over the last 50 years, it still retains a mortality rate of 40–85%, depending on the clinical background of the patient [1,2]. Occurring primarily in the hospitalized patient, Gram-negative sepsis has risen from 0.8% of all hospital admissions in 1958 to greater than 1.0% currently [3]. Overall, shock is seen as an acute syndrome of the cardiovascular system in which the microcirculation is deranged so that critical organs of the body are inadequately perfused [4]. Many biochemical alterations are seen at the cellular level which contribute to the lesion as a whole. Endotoxin is known to produce many of these alterations directly or indirectly through the involvement of endogenous mediators [5–7]. Inhibition of fundamental biochemical pathways relatively early in endotoxemia may contribute to the eventual irreversibility of the shock syndrome.

Carbohydrate Loss During Endotoxic Shock

Two decades ago, employing the murine typhoid model, Berry and Smythe [8] were the first to show a correlation between sugar loss and endotoxin release during Gram-negative septicemia. Carbohydrate loss during endotoxic shock has been recognized for many years; however, the mechanisms responsible for the hypoglycemia associated with this syndrome are only now being elucidated. It is quite clear from the work of many investigators that not only does a hypoglycemic state exist in shock, but stored sugar compounds, eg, glycogen, are rapidly depleted as well. This can be seen in Figure 1, in which liver glucose and glycogen levels were measured at intervals after treatment in mice given an LD_{50} (median lethal dose) of endotoxin extracted from *Salmonella typhimurium* by hot aqueous phenol. Although liver glucose levels were not lower than control

Pathophysiological Effects of Endotoxins at the Cellular Level, pages 99–113

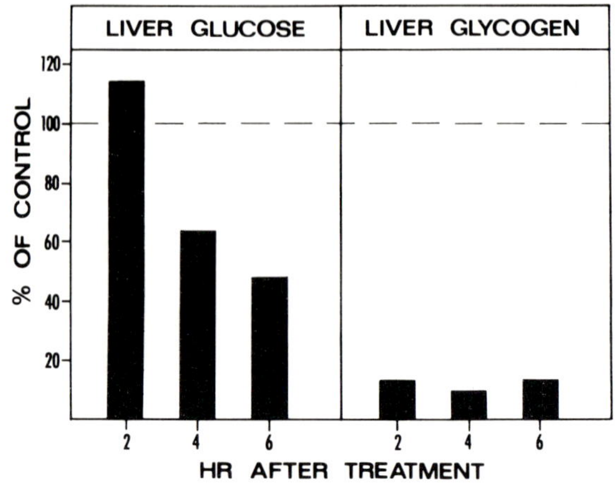

Fig. 1. Effect of endotoxin (LD_{50}) on hepatic glucose and glycogen levels in mice at 2, 4, and 6 hr after treatment. Animals were injected ip with endotoxin from *Salmonella typhimurium*, and liver samples were obtained at the times indicated.

values until 4 hr, liver glycogen was significantly reduced by 2 hr and remained at minimal levels. In these animals an initial hyperglycemia was seen, followed quickly by a drop in blood glucose levels.

Inhibition of Hepatic Glycogen Synthesis and Gluconeogenesis

Although enhanced glycogenolysis [9] and stimulated glycolysis [10] have been suggested as responsible for carbohydrate loss during shock, evidence from our laboratory implicates that lesions in carbohydrate synthesis are primarily involved. The complexity of hepatic carbohydrate metabolism is depicted in Figure 2, and it can be seen that certain rate-limiting steps in carbohydrate catabolism are not fully reversible. In the resynthesis of sugar, it is these steps that must be circumvented. Interestingly, endotoxin treatment appears to act at these steps, specifically affecting enzymes critical for the synthesis of glucose from nonsugar sources. McCallum and Berry [11] demonstrated inhibition of hepatic glucose-6-phosphatase and fructose-1, 6-diphosphatase to occur relatively late after endotoxin treatment. In a more detailed study of enzymes of the glycogen cycle, Giger and McCallum [12] found that endotoxin treatment did not alter total glycogen synthase activity, but the amount of the enzyme in the active form was significantly lower in endotoxic mice as early as 4 hr after treatment. Liver glycogen phosphorylase was not significantly activated in endotoxin-poisoned mice, contraindicating enhanced glycogenolysis as a mechanism for depletion of carbohydrate during shock. Altered activation of glycogen synthase, however, most surely contributes to the loss of glycogen during endotoxemia.

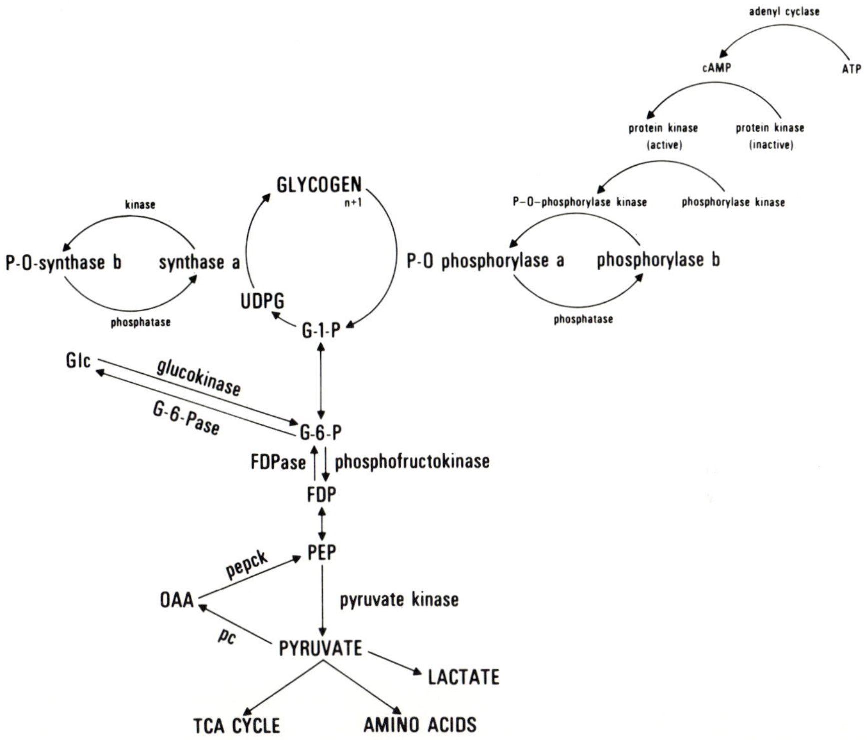

Fig. 2. Pathways of hepatic glucose and glycogen metabolism. Endotoxin treatment (LD_{50}) has been shown to specifically affect rate-limiting enzymes: Glycogen synthase a [12,13], glucose-6-phosphatase [11], fructose 1,6-diphosphatase [11], phosphoenolpyruvate carboxykinase [18,19], and pyruvate kinase [10].

Of the many metabolic effects produced by bacterial endotoxin, inhibition of hepatic gluconeogenesis is perhaps the most significant because it is the key pathway for glucose synthesis in animals under stress [7]. A demonstrable inhibition of hepatic gluconeogenesis was seen in endotoxic mice by McCallum and Berry [13]. As early as 4 hr after treatment with an LD_{50}, the incorporation of the ^{14}C-label from ^{14}C(U)-alanine into blood glucose was reduced to approximately 50% of the amount seen in control animals. Other investigators [14–17] have noted impaired gluconeogenic and glycolytic flux in poisoned animals. Inhibition of hepatic phosphoenolpyruvate carboxykinase (PEPCK), a critical enzyme in gluconeogenesis, has been shown to occur in poisoned mice [18,19].

Altered Metabolism in Isolated Hepatocytes

Filkins and Cornell [20] found that endotoxin administration depressed maximal gluconeogenic rates when evaluated in vitro with isolated rat hepatocytes.

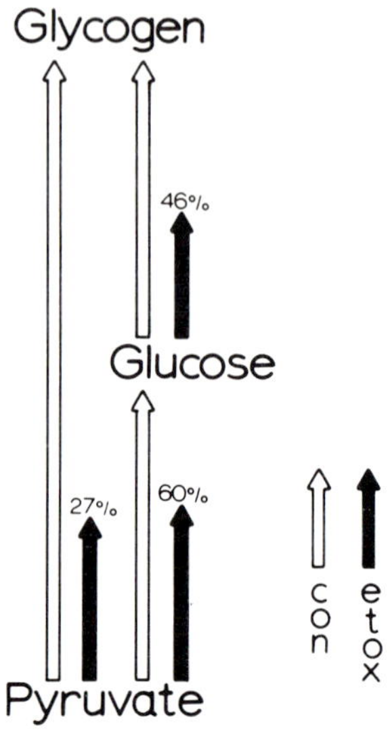

Fig. 3. Extent of inhibition of gluconeogenesis (pyruvate → glucose), glyconeogenesis (pyruvate → glycogen), and glycogenesis (glucose → glycogen) observed in rat hepatocytes isolated 6 hr after endotoxin treatment (LD_{50}). The percentage of control value is indicated above each shaded bar. Con, control; Etox, endotoxin treated.

They found endotoxin did not alter gluconeogenesis, however, when it was added directly to hepatocytes obtained from normal fasted rats. In our laboratory, inhibition of gluconeogenesis from pyruvate was also seen in hepatocytes isolated from poisoned rats [RE McCallum, Abstr Ann Meet Am Soc Microbiol, 1975, B32, p 16]. As can be seen in Figure 3, the incorporation of radiolabeled pyruvate and glucose into glycogen in liver parenchymal cells from endotoxin-treated animals was 27 and 46% of the control values, respectively. Measurements of PEPCK indicated that the enzyme activity in hepatocytes from poisoned rats was less than half of the control values. Furthermore, the direct addition of endotoxin to normal rat liver parenchymal cells was without effect on either gluconeogenesis or PEPCK activity.

Mediated Aspects of Hepatic PEPCK Inhibition

The results obtained using isolated hepatocytes suggested the possible participation of soluble mediators in the inhibition of hepatocyte gluconeogenesis during endotoxic shock. The involvement of mediators in a wide array of biological effects of endotoxin has received considerable attention [6,7] and was the subject of a recent symposium [5]. Moore et al [21] demonstrated that the inhibition of glucocorticoid induction of hepatic PEPCK was a mediated event.

The involvement of the host reticuloendothelial system (RES) in production of this mediator was suggested because macrophages cultured in the presence of endotoxin yielded a supernatant fluid that blocked PEPCK induction in tolerant (unresponsive) animals. This glucocorticoid antagonizing factor (GAF) was subsequently shown not to be produced in low-responder (C_3H/HeJ) mice [22,23]. We recently investigated the carbohydrate-depleting capability of endotoxin in C_3H/HeJ mice [24]. The transfer of peritoneal cells from high-responder (C_3HeB/FeJ) mice to low-responder mice appeared to sensitize the recipients to endotoxin-induced carbohydrate loss. Furthermore, the transfer of plasma from low responders seemed to protect the high-responder recipients from sugar loss. These results also tend to support the involvement of mediators in carbohydrate depletion seen during shock.

Significant involvement of mediators or host-modified endotoxin, or both, in inhibition of hepatic gluconeogenesis by endotoxin was the subject of a recent report from this laboratory [25]. Inhibition of hepatocyte PEPCK activity in a hepatocyte–Kupffer cell system in vitro is summarized in Table I. It can be seen that typical inhibition of PEPCK was not seen when *S. typhimurium* endotoxin was added directly to normal liver homogenates, or to normal hepatocytes. Serum from endotoxin-poisoned mice displayed only slight inhibitory activity, whereas serum from endotoxic mice pretreated with zymosan was active in blocking PEPCK activity in normal hepatocytes. The addition of culture supernatant fluids from Kupffer cells incubated with 10 μg of endotoxin for 2 hr to normal hepatocytes did cause a slight but significant decrease in both gluconeogenic capability and PEPCK activity. These results suggest that Kupffer cells may be the source of mediator(s) (or macrophage-modified endotoxin) responsible for inhibition of gluconeogenesis in hepatic parenchymal cells. Results from experiments employing Kupffer cells from mice pretreated with zymosan suggest the "activated" macrophage may release higher amounts of putative inhibitory material than nonpretreated controls.

The present study was undertaken in order to evaluate hepatocyte–Kupffer cell interactions in the inhibition of hepatic gluconeogenesis by endotoxin. Employing high-responder C_3HeB/FeJ and low-responder C_3H/HeJ mouse strains as sources for hepatocytes and Kupffer cells, it was hoped that the mediated basis for this inhibition might become more clear.

MATERIALS AND METHODS

Animals

Female inbred mice (high-responder C_3HeB/FeJ and low-responder C_3H/HeJ strains) were purchased from the Jackson Laboratory (Bar Harbor, ME). Mice were housed in sterile cages containing sterile bedding in a laminar flow hood. They were allowed food (sterilized Purina Lab Chow) and water (pH 2.5, con-

TABLE I. Summary of the Mediated Aspects of the Inhibition of Hepatic Phosphoenolpyruvate Carboxykinase (PEPCK) Activity in Mice Poisoned with *Salmonella typhimurium* Endotoxin (modified from [25])

Treatment	Effect on hepatocyte PEPCK enzyme activity
1. Whole liver homogenates from endotoxin[a]-poisoned mice	Decrease[f]
2. Hepatocytes isolated from endotoxin[a]-poisoned mice	Decrease[f]
3. Normal liver homogenates + endotoxin[b] added in vitro	No change
4. Normal hepatocyte homogenates + endotoxin[b] added in vitro	No change
5. Normal intact hepatocytes[c] + endotoxin[b] added in vitro	No change
6. Same as 5 + serum from endotoxin[a]-poisoned mice	
a. Incubated at 37°C, 15 min	No change
b. Incubated at 37°C, 60 min	Decrease[f]
7. Same as 5 + serum from zymosan[d]-pretreated mice given endotoxin	Decrease[f]
8. Same as 5 + culture fluid[e] from normal Kupffer cells	No change
9. Same as 5 + culture fluid[e] from Kupffer cells incubated with endotoxin (10 μg) for 2 hr	Decrease[f]

[a]Six hours after 1 LD_{50}, injected intraperitoneally.
[b]Concentration range of 0.01 to 100 μg.
[c]Approximately 5×10^6 to 1×10^7 hepatocytes in Krebs–Ringer bicarbonate buffer (pH 7.4, containing 5% defatted bovine serum albumin and gassed with 95% 0_2–5% CO_2) were incubated with equal volumes of the materials listed in a total volume of 3 ml. After 30 min at 37°C, except where noted, hepatocyte suspensions were homogenized, and enzyme activity was determined in supernatant fluids resulting from centrifugation at 20,000g.
[d]Zymosan (0.5 mg/0.1 ml) was given intravenously, followed 48 hr later with 1.0 mg in 0.5 ml, intraperitoneally. Five days later 25 μg of endotoxin was given intravenously, and 6 hr later the resulting serum was added to normal hepatocytes.
[e]After adherence to plastic petri dishes and washing, 3×10^6 normal Kupffer cells in monolayer were incubated for 24 hrs with Eagle minimal essential medium containing L-glutamine, 10% fetal calf serum, and antibiotics. The culture medium was replaced and after an additional 2 hr incubation, the entire culture fluid (2 ml) was added to approximately 1×10^7 normal mouse hepatocytes.
[f]Significantly less than control values ($P \leqq 0.05$).

taining menadione supplement, 0.5 μg/ml) ad libitum. All animals were fasted overnight prior to the beginning of each experiment at 0800 hr.

Bacterial Strains and Endotoxins

All bacteria were grown to late exponential growth phase in minimal E Medium containing 0.5% glucose [26]. RNA- and protein-free lipopolysaccharide was prepared from *Salmonella typhimurium* strain SR-11 by the hot aqueous phenol method of Westphal and Jann [27]. Except where specifically noted, this type of endotoxin was used in all experiments. Boivin extracts of *Escherichia coli* 0111:B4 were prepared with cold tricholoroacetic acid [28] and R_e glycolipid was extracted from *Salmonella minnesota* strain R595 by the Galanos method [29]. In all animal experiments, endotoxin was injected by the intraperitoneal route.

Hepatocyte and Kupffer Cell Isolation and Cell Culture

Mouse hepatocytes and Kupffer cells were isolated from perfused mouse liver, as described previously [25]. Briefly, parenchymal cells were obtained by reversed organ perfusion in vitro with 50 mg% collagenase in Krebs–Ringer bicarbonate buffer. Cell viability was monitored by exclusion of trypan blue dye (preparations with less than 85% viable cells were not used).

Nonparenchymal cells (hereafter referred to as Kupffer cells) were isolated from mouse liver by perfusion in vitro with 0.2% Pronase and 1000 U of DNAse according to the procedure of Emeis and Planque [30]. These cells were judged to be Kupffer cells based on their ability to phagocytize carbon particles and live *Candida pseudotropicalis*. These preparations contained less than 1–2% endothelial and fibroblastic cells (by phase microscopy). Following adherence to plastic petri dishes, Kupffer cell monolayers were cultured 24 hr prior to the addition of 10 μg of endotoxin. These cells were cultured in the presence of endotoxin for 2 hr and then the culture supernatant fluids were added to 1×10^7 normal mouse hepatocytes. After 60 min of incubation at 37°C, the hepatocyte suspensions were homogenized and PEPCK activity was determined.

PEPCK Enzyme Assay

Phosphoenolpyruvate carboxykinase (PEPCK) activity was determined in supernatant fluids (20,000g) of hepatocyte homogenates by following the condensation of $^{14}CO_2$ with phosphoenolpyruvate (PEP) to form oxalacetate, according to the method of Chang and Lane [31]. Activity is expressed as nanomoles of oxalacetic acid formed per milligram of cell protein per hour.

Statistics

Mean values obtained from several experiments were analyzed by Rank order correlation and Student's t test for significant differences ($P \leq 0.05$) [32].

TABLE II. Inhibition of Hepatocyte Phosphoenolpyruvate Carboxykinase (PEPCK) Activity by Endotoxin in C_3HeB/FeJ versus C_3H/HeJ Mouse Strains (6 hr after intraperitoneal administration of 100 μg)

Mouse strain and treatment	Hepatocyte PEPCK activity (units)
C_3HeB/FeJ control	85.2 ± 4.4[a]
C_3HeB/FeJ + endotoxin	43.8 ± 3.2[b]
C_3H/HeJ control	93.8 ± 5.1
C_3H/HeJ + endotoxin	87.4 ± 4.1

[a]Mean value ± SEM from triplicate determinations on three separate hepatocyte preparations.
[b]Significantly lower than control values ($P \leqq 0.05$).

RESULTS

Comparison of C_3HeB/FeJ and C_3H/HeJ Mouse Strains

PEPCK activity in hepatocytes isolated from poisoned mice. We previously demonstrated that the carbohydrate depletion seen in high-responder C_3HeB/FeJ mice given an LD_{50} dose of *S. typhimurium* endotoxin did not occur in low-responder C_3H/HeJ mice given the same dose [24]. For this reason, PEPCK activity was measured in hepatocytes isolated from both C_3HeB/FeJ and C_3H/HeJ mice 6 hr after 100 μg of endotoxin was administered. The results are seen in Table II. It can be seen that enzyme activity dropped to approximately 50% of the nonpoisoned control values in C_3HeB/FeJ mice. This is a characteristic response and corresponds to previous results [24,25]. Hepatocytes obtained from C_3H/HeJ mice, however, did not demonstrate a decrease in PEPCK activity 6 hr after endotoxin treatment. This further substantiates the relative unresponsiveness of the C_3H/HeJ mouse to endotoxin.

In vitro experiments—role of the macrophage. Since previous results suggested possible involvement of soluble mediators or macrophage-modified endotoxin in the inhibition of hepatocyte PEPCK activity, experiments were designed to pursue this aspect in C_3HeB/FeJ versus C_3H/HeJ mice. The results seen in Table III indicate that Kupffer cells isolated from normal C_3HeB/FeJ mice and incubated in vitro with endotoxin yielded an inhibitor of hepatocyte PEPCK not obtained when Kupffer cells from C_3H/HeJ mice were employed. It is apparent that the source (mouse strain) of the hepatocytes was not critical, since a decrease in enzyme activity was seen when supernatant fluids from endotoxin–C_3HeB/FeJ Kupffer cells were mixed with either C_3HeB/FeJ or C_3H/HeJ hepatocytes. This indicates that hepatocyte PEPCK activity is decreased in cells from unresponsive mice when inhibitory substances resulting from the

TABLE III. Inhibition of Hepatocyte Phosphoenolpyruvate Carboxykinase (PEPCK) Activity by Supernatant Fluids* from Kupffer Cell Cultures Incubated with *S. typhimurium* Endotoxin

Mouse strain employed		Hepatocyte PEPCK activity (units)
Kupffer cells	Hepatocytes	
C_3HeB/FeJ	C_3HeB/FeJ	$14.9 \pm 2.1^{a,b}$
C_3HeB/FeJ	C_3H/HeJ	16.0 ± 1.8^{b}
C_3H/HeJ	C_3HeB/FeJ	21.9 ± 1.6
C_3H/HeJ	C_3H/HeJ	20.3 ± 2.0

*See footnote e in Table I.
[a]Mean value ± SEM of triplicate determinations from three separate experiments.
[b]Significantly lower than values obtained with C_3H/HeJ Kupffer cells ($P \leqq 0.05$).

interaction of normally responsive Kupffer cells (macrophages) and endotoxin are added to them. This effect is not seen when endotoxin is added directly to hepatocytes from responsive mice (Table I).

Employing Kupffer cells from C_3H/HeJ mice, no significant difference in hepatocyte PEPCK activity from controls was seen following the addition of supernatant fluids from macrophage–endotoxin mixtures. These results implicate the Kupffer cell of the normally responsive mouse as the source of either modified endotoxin or soluble mediator which inhibits hepatocyte PEPCK activity. As was reported previously [25], this observed inhibitory effect may involve the soluble mediator, glucocorticoid antagonizing factor (GAF), reported by Moore et al [21] or perhaps another material, since direct inhibition of existing PEPCK activity was seen.

Variable Response in Hepatocyte PEPCK Due to Type of Endotoxin Employed

Another important aspect of the metabolic responses seen in endotoxemia is the type of endotoxic material used to imitate experimentally the conditions that actually occur in septic shock. This point has received recent emphasis by Morrison and Ulevitch [6]. For this reason, we examined the possible effects of various endotoxin preparations on hepatocyte PEPCK activity when endotoxin and C_3H/HeJ (low-responder) Kupffer cells were incubated together. The results of these experiments are presented in Table IV. When high-responder C_3HeB/FeJ Kupffer cell culture fluids were used, hepatocyte PEPCK activity was decreased to approximately 75% of control values regardless of the source of endotoxin or extraction method used (data not shown). As can be seen in Table IV, the use of phenol-extracted endotoxin from *Salmonella typhimurium* or PCP-extracted glycolipid from *Salmonella minnesota* did not cause a significant decrease in hepatocyte PEPCK activity when extracts were incubated with $C_3H/$

TABLE IV. Inhibition of C_3H/HeJ Mouse Hepatocyte Phosphoenolpyruvate Carboxykinase (PEPCK) by Culture Fluids* from C_3H/HeJ Mouse Kupffer Cells Incubated with Various Endotoxin Preparations

Source of endotoxin and preparation	Hepatocyte PEPCK activity (units)
Control cultures without endotoxins added	22.5 ± 2.1[a]
Salmonella typhimurium SR-11, hot phenol [27]	21.8 ± 1.6
Salmonella minnesota R595, PCP [29]	23.1 ± 2.0
Escherichia coli 0111:B4, cold TCA [28]	17.2 ± 1.6[b]

*See footnote e in Table I.
[a]Mean value ± SEM obtained from duplicate determinations from two separate experiments.
[b]Significantly lower than values obtained with *S. typhimurium* and *S. minnesota* endotoxin ($P \leqq 0.05$).

HeJ Kupffer cells. These results are consistent with data presented in Table III and those reported previously [25]. The most interesting result in Table IV is the apparent decrease in hepatocyte PEPCK activity (76% of the control value) when TCA-extracted endotoxin from *E. coli* was incubated with C_3H/HeJ Kupffer cells. These results indicate that components not present in conventional lipopolysaccharide preparations, but found in Boivin extracts, may elicit a response in normally unresponsive C_3H/HeJ mice. This observation has been reported by others [6,22,23] and, yet, in the case of altered hepatic PEPCK activity, remains to be explained.

DISCUSSION

Infections caused by Gram-negative bacilli have become increasingly prevalent and constitute the most frequent type of nosocomial infection [33]. The often cited statistics of McCabe et al [34] suggest that greater than 300,000 episodes of Gram-negative septic shock with more than 100,000 fatalities may occur each year in the United States. The premise upon which the efforts in our laboratory have been based is that soon (less than 6 hr after poisoning) after bacterial endotoxin is given to experimental animals, subtle, yet critical, biochemical events occur which eventually contribute to the irreversibility of the shock syndrome. We have shown that the carbohydrate depletion that occurs in experimental endotoxemia is rapid and irreversible. A thorough understanding of the mechanisms involved in this fundamental response may suggest a means by which reversal can be brought about. Since there is a good possibility that endotoxin may elicit its effects on host metabolism via mediators (or some type of host-modified endotoxin), complex metabolic pathways may be a vulnerable target. Further study of the specific metabolic lesions involved may lead to elucidation of a "common mediator" of endotoxin actions.

The significance of metabolic homeostasis in an animal's ability to overcome stress cannot be overemphasized. As mentioned by Beisel [35], infectious processes usually elicit a complex array of metabolic responses within a host. Since alterations in carbohydrate metabolism have been demonstrated in such a wide variety of infections, intoxications, and other disorders, disruption of the fine balance of homeostasis may be one of the key events necessary for the above conditions to develop. Hence, our justification for continued study of metabolic changes occurring at the cellular level during experimental endotoxic shock.

Models of endotoxic shock in laboratory animals demonstrate many of the pathophysiological alterations seen in septic shock in man, although considerable variation in some responses is seen in different species. This variation has been found to be a useful tool for analyzing the mechanisms of endotoxin action at the cellular level. The C_3H/HeJ mouse strain has been found to be resistant to many of the biological effects of endotoxin [5,36,37]. As a result, this strain has been widely used in recent years to study the interaction of endotoxin with mammalian cells. Subsequently, many of the responses to endotoxin either altered or lacking in the low responder mouse have been found to be under genetic control [38,39]. It is clear from the evidence given that the defect in endotoxin-induced stimulation in the C_3H/HeJ mouse is related directly to defective responses of their lymphoid cells. Work by Rosenstreich et al [40] and others [5] suggests the cell type most likely to be involved is the macrophage.

The results of the current study shed further light on the mechanisms involved in the inhibition of hepatocyte PEPCK activity, a critical lesion in the perturbed synthesis of sugars from amino acids in livers of endotoxin-poisoned animals. It was shown that hepatocyte PEPCK activity in cells isolated from C_3H/HeJ mice 6 hr after endotoxin was not different from control values (Table II). A conventional LD_{50} dose (5 mg/kg) for C_3HeB/FeJ (high-responder) mice was used in these experiments. The LD_{50} for low-responder mice is approximately 55 mg/kg; thus, it can be seen that C_3H/HeJ mice are relatively unresponsive to the inhibition of PEPCK or gluconeogenesis by endotoxin. For this reason this particular mouse strain can be used to study the involvement of mediators or host-modified endotoxin in the altered metabolism seen in endotoxic shock.

A role for the macrophage as a target cell or possible source of mediators in endotoxic shock has been suggested by a number of investigators [5]. Evidence presented in Table III confirms the relative endotoxin unresponsiveness of liver macrophages (Kupffer cells) from C_3H/HeJ mice. Of significance is the observation that PEPCK activity in either C_3HeB/FeJ hepatocytes or C_3H/HeJ hepatocytes was decreased when supernatant fluids from C_3HeB/FeJ Kupffer

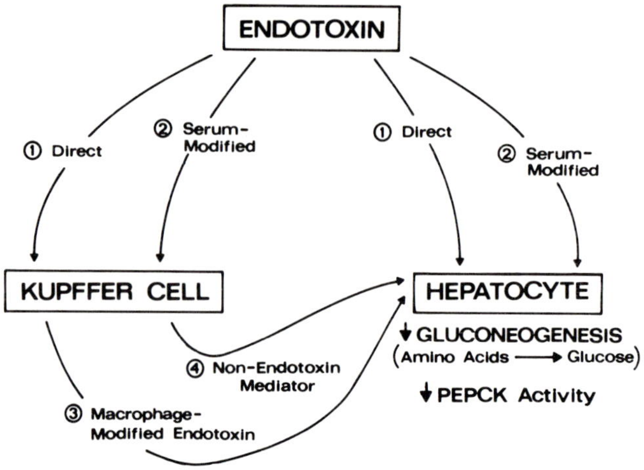

Fig. 4. Hepatocyte–Kupffer cell interactions possibly involved in endotoxin inhibition of hepatic gluconeogenesis. At least four possible pathways may account for decreased hepatic phosphoenolpyruvate carboxykinase (PEPCK) activity seen in endotoxin shock. Although not completely ruled out by the present study, a direct action of endotoxin (*1*) is not likely.

cell–endotoxin mixtures were used. In other words, PEPCK activity in C_3H/HeJ hepatocytes is fully responsive to the inhibitory effects of endotoxin, but not directly. This result places considerable emphasis on macrophages as a source of this inhibition.

There are, of course, a myriad of possible hepatocyte–Kupffer cell interactions which occur in endotoxin-treated animals. A simplified scheme is presented in Figure 4. According to these pathways, endotoxin treatment may effect decreased hepatocyte PEPCK activity in at least four different ways: A direct inhibition of the enzyme by intact endotoxin *1* (Fig. 4) is not likely in light of results of our recent studies [24,25] summarized in Table I. A direct effect of endotoxin on Kupffer cells, however, is quite possible—with a resulting decrease in enzyme activity due to macrophage–endotoxin interaction. Endotoxin may be modified by the host via several possible ways (eg, binding to serum components, serum-induced detoxification, or cell-mediated degradation or modification). The possibility that serum-modified endotoxin *2* may act either directly on hepatocytes, or indirectly through Kupffer cells, is not ruled out by the results of the present study. In fact a recent report by Ulevitch et al suggests serum lipoprotein–endotoxin complexes may possess biological activity ([41] see also this volume). Further work looking at the effect of these complexes applied directly on isolated hepatocytes may determine this to be a fundamental mechanism. The possibility that macrophages (Kupffer cells) may modify endotoxin *3* to the extent that low-responder mice suddenly become fully responsive has not been examined, but should also be considered.

Most attention has recently focused on the participation of endogenous mediators 4 in the changes seen during endotoxin shock [5]. The results presented here are consistent with the concept, yet a specific "inhibitor" of hepatocyte PEPCK has not yet been isolated. The mediator GAF [42] may be the factor involved, but the primary effect of this inhibitor is to block the glucocorticoid-induced synthesis of PEPCK [21,23]. Although the inhibition of PEPCK reported in this study was not complete (about 69% of control values) this amount of inhibition of an enzyme which functions at a rate-limiting step in a metabolic pathway is critical, ie, the catalytic potential of the enzyme is dramatically reduced. Whatever the pathway(s) for hepatocyte–Kupffer cell interaction in the inhibition of PEPCK, it is certain to be complex. It seems likely that more than one of the pathways presented in Figure 4. may occur simultaneously. Thus, one might anticipate serum lipoprotein–endotoxin complexes reacting with Kupffer cells, hepatocytes, or both—while soluble mediators released from macrophages may also be affecting hepatocytes. The use of low-responder (C_3H/HeJ) mice in conjunction with syngeneic normal-responder (C_3HeB/FeJ) mice affords a complex, yet interesting, system with which to pursue these questions.

Mention should be made of the observation in this study that hepatocyte-inhibitory substance(s) was (were) released by Kupffer cells from low-responder mice when mixed with endotoxin extracted with cold tricholoracetic acid. This effect was not seen when endotoxin extracted from a smooth bacterium with hot aqueous phenol or endotoxin extracted from a rough organism with phenol–choloroform–petroleum ether was used. The TCA method is known to extract material (protein) other than lipopolysaccharide from Gram-negative bacterial cell walls. Whether or not endotoxin protein (EP) [43] or lipid A-associated protein (LAP) [44] is responsible for this inhibitory effect is currently under investigation. When one designs studies of the possible role of endotoxin in Gram-negative sepsis, crude extracts should probably be used because the bacterial cell wall fragments that circulate in septic patients are certainly not highly purified. In this way, perhaps, a more thorough understanding of the role of endotoxin in the pathogenesis of Gram-negative sepsis will be reached.

ACKNOWLEDGMENTS

This study was supported, in part, by U.S. Public Health Service Grant AI11238 from the National Institute of Allergy and Infectious Diseases and a grant from the Research Council of the Faculty Senate of the University of Oklahoma Health Sciences Center.

The excellent technical assistance of Hilary Thirkill is gratefully acknowledged.

REFERENCES

1. Ledingham IMcA,Sleigh JD:"Shock:Clinical and Experimental Aspects." New York: Exc Med, 1976, p 273.

2. Lillehei RC,Dietzmen RH:"Priniciples of Surgery." New York: McGraw–Hill,1974,p 133.
3. von Graevenitz A: Gram-negative rods as agents of nosocomal disease: Some recent developments with comments on mixed cultures.In Urbaschek B,Urbaschek R, Neter E(eds):"Gram-Negative Bacterial Infections and Mode of Endotoxin Actions." Vienna: Springer–Verlag, 1975,p 30.
4. Urbaschek B, Urbaschek R,Neter E:"Gram-Negative Bacterial Infections and Mode of Endotoxin Actions." Vienna: Springer–Verlag,1975.
5. Mergenhagen S,Berry LJ, Rosenstreich D:Endogenous mediators in host responses to bacterial endotoxin. In Schlessinger D(ed):"Microbiology,1980." Washington,DC: American Society for Microbiology, 1980,pp 3–167.
6. Morrison DC,Ulevitch RJ: The effects of bacterial endotoxins on host mediation systems. Am J Pathol 93:525,1978.
7. Berry LJ: Bacterial toxins.Crit Rev Toxicol 5:239,1977.
8. Berry LJ,Smythe DS: Some metabolic aspects of host–parasite interactions in the mouse typhoid model. Ann NY Acad Sci 88:1278,1960.
9. Hamosh M,Shapiro B:The mechanism of glycogenolytic action of endotoxin. Br J Exp Pathol 41:372,1960.
10. Smith SM,Snyder IS:Effect of lipopolysaccharide and lipid A on mouse liver pyruvate kinase activity. Infect Immun 12:993,1975.
11. McCallum RE,Berry LJ: Mouse liver fructose-1,6-diphosphatase and glucose-6-phosphatase activities after endotoxin poisoning.Infect Immun 6:883,1972.
12. Giger O,McCallum RE: Response of mouse liver glycogen cycle enzymes to endotoxin treatment. Am J Physiol 231:1285,1976.
13. McCallum RE,Berry LJ: Effects of endotoxin on gluconeogenesis,glycogen synthesis, and liver glycogen synthase in mice.Infect Immun 7:642,1973.
14. Urbaschek B, Urbaschek R: Some aspects of microcirculatory and metabolic changes in endotoxemia and in endotoxin tolerance. In Schlessinger D(ed):"Microbiology, 1977." Washington DC: American Society for Microbiology,1977,p 286.
15. Williamson JR,Refino C,LaNoue K: Effects of *E. coli* lipopolysaccharide B treatment of rats on gluconeogenesis.In Porter R,Knight J(eds):"Energy Metabolism in Trauma." London: J&A Churchill,1970,p 145.
16. Schumer W: Cellular metabolism in shock.Circ Shock 2:109,1975.
17. Shands JW Jr,Miller V,Martin H,Senterfitt V: Hypoglycemic activity of endotoxin II Mechanism of the phenomenon in BCG-infected mice. J Bacteriol 98:494,1969.
18. Rippe DF,Berry LJ: Study of inhibition of induction of phosphoenolpyruvate carboxykinase by endotoxin by radial immunodiffusion. Infect Immun 6:766,1972.
19. Berry LJ,Smythe DS,Colwell LS: Inhibition of hepatic enzyme induction as a sensitive assay for endotoxin. J Bacteriol 96:91,1968.
20. Filkins JP,Cornell RP: Depression of hepatic gluconeogenesis and the hypoglycemia of endotoxin shock. Am J Physiol 227:778,1974.
21. Moore RN,Goodrum KJ,Berry LJ: Mediation of endotoxin effects by macrophages.J Reticuloendothel Soc 19:187,1976.
22. Moore RN,Goodrum KJ,Couch RE Jr,Berry LJ: Factors affecting macrophage function:Glucocorticoid antagonizing factor.J Reticuloendothel Soc 23:321,1978.
23. Moore RN,Goodrum KJ,Couch RE Jr,Berry LJ: Elicitation of endotoxemic effects in C_3H/HeJ mice with glucocorticoid antagonizing factor and partial characterization of the factor.Infect Immun 19:79,1978.
24. McCallum RE,Pontious ML:Carbohydrate metabolism in C_3H/HeJ (nonresponder) mice during endotoxic shock. Advan Shock Res 2:113,1979.

25. McCallum RE: Mediated inhibition of hepatic gluconeogenesis by endotoxin.In Schlessinger D (ed):"Microbiology,1980." Washington DC: American Society for Microbiology,1980,p 87.
26. Vogel HJ,Bonner DM: Acetylornithinase of *Escherichia coli:* Partial purification and some properties.J Biol Chem 218:97,1956.
27. Westphal O,Jann K:Bacterial lipopolysaccharides: Extraction with phenol water and further applications of the procedure.In Whistler RL (ed):"Methods in Carbohydrate Chemistry." Vol 5. New York: Academic Press,1965,p 83.
28. Boivin A,Mesrobeanu L: Recherches sur les antigènes somatique et sur les endotoxines des bactéries. I Considérations générales et exposé des techniques utiliseés. Rev Immunol (Paris) 1:553,1935.
29. Galanos C,Lüderitz O,Westphal O: A new method for the extraction of R lipopolysaccharides. Eur J Biochem 9:245,1969.
30. Emeis JJ,Planque B: Heterogeneity of cells isolated from rat liver by pronase digestion: Ultrastructure cytochemistry and cell culture.J Reticuloendothel Soc 20:11,1976.
31. Chang HC,Lane MD: The enzymatic carboxylation of phosphoenolpyruvate II Purification and properties of liver mitochondrial phosphoenolpyruvate carboxykinase. J Biol Chem 241:2413,1966.
32. Lewis AE: "Biostatistics." New York: Rheinhold,1966, p 55.
33. Neter E: The nature of specific and nonspecific Gram-negative infections including enteropathogenic *E. coli* enteritis.In Urbaschek B,Urbaschek R,Neter E (eds): "Gram-Negative Bacterial Infections and Mode of Endotoxin Actions." Vienna: Springer–Verlag,1975,p 3.
34. McCabe WR,Kreger BE,Johns M: Type-specific and cross-reactive antibodies in Gram-negative bacteremia. N Eng J Med 287:261,1972.
35. Beisel WR: Magnitude of the host nutritional responses to infection. Am J Clin Nutr 30:1236,1977.
36. Sultzer BM: Genetic control of host responses to endotoxin. Infect Immun 5:107,1972.
37. Sultzer BM,Goodman GW: Characteristics of endotoxin-resistant low-responder mice.In Schelssinger D (ed): "Microbiology,1977." Washington DC: American Society for Microbiology, 1977,p 304.
38. Watson J,Kelly K,Largen M: Genetic and cellular aspects of host response to endotoxin. In Schlessinger D (ed):"Microbiology,1977." Washington DC: American Society for Microbiology,1977,p 298.
39. Watson J,Kelly K,Largen M,Taylor B: The genetic mapping of a defective LPS response gene in C_3H/HeJ mice. J Immunol 120:422,1978.
40. Rosenstreich DL,Glode LM, Wahl LM,Sandberg AL, Mergenhagen SE: Analysis of the cellular defects of endotoxin-unresponsive C_3H/HeJ mice.In Schlessinger D (ed):"Microbiology,1977." Washington DC: American Society for Microbiology,1977,p 314.
41. Ulevitch RJ,Johnston AR, Weinstein DB: New function for high density lipoproteins.Their participation in intravascular reactions of bacterial lipopolysaccharides. J Clin Invest 64:1516,1979.
42. Berry LJ, Goodrum KJ,Ford CW,Resnick IG,Shackleford GM: Partial characterization of glucocorticoid antagonizing factor in hepatoma cells. In Schlessinger D (ed):"Microbiology,1980."Washington DC: American Society for Microbiology, 1980,p 77.
43. Sultzer BM,Goodman GW: Endotoxin protein: A B-cell mitogen and polyclonal activator of C_3H/HeJ lymphocytes. J Exp Med 144:821,1976.
44. Morrison DC,Betz SJ,Jacobs DM: Isolation of lipid A-bound polypeptide responsible for "LPS-initiated" mitogenesis of C_3H/HeJ spleen cells. J Exp Med 144:840,1976.

Influence of *Escherichia coli* Endotoxin on Palmitate, Glucose, and Lactate Utilization by Isolated Dog Heart Myocytes

Maw-Shung Liu, Will M. Long, and John J. Spitzer

We have previously reported that endotoxin alters myocardial substrate utilization both under in vivo and in vitro conditions [1,2]. Following in vivo administration of endotoxin [1], palmitate oxidation by heart homogenates was decreased and lactate oxidation was increased. Unlike in the in vivo studies, oxidations of palmitate and lactate were both decreased in heart homogenates incubated with endotoxin in vitro [2]. The decreases in substrate oxidation in vitro were associated with the release of Ca^{2+} ions from the homogenates [2]. In order to investigate further the in vitro effects of endotoxin, we reexamined the influence of endotoxin on myocardial substrate utilization by using isolated beating myocytes. Since isolated myocytes appeared to possess functionally intact plasma membranes [3], this preparation also offered an opportunity for the study of the endotoxin effect on metabolite transport.

MATERIALS AND METHODS

Mongrel dogs of either sex weighing 15–20 kg were fasted overnight but were permitted water prior to the experiments. Following anesthesia with 2% chloralose in 20% urethane, a thoracotomy was performed, and the heart was excised and placed immediately into chilled, Ca^{2+}-free phosphate buffer. Heart myocytes were isolated from ventricles as described earlier [3]. The isolation procedure consisted of a series of repeated enzymatic digestions with collagenase and hyaluronidase, mechanical shaking, and Ficoll centrifugation. About 80% of isolated myocytes excluded trypan blue. Myocytes which excluded trypan blue were contracting autorhythmically. The isolated cells were suspended in a Ca^{2+}-free Krebs–Ringer's solution (pH 7.4) for the subsequent metabolic studies.

Pathophysiological Effects of Endotoxins at the Cellular Level, pages 115 – 121

Oxidation of palmitate by cardiac muscle cells was performed as described previously [3] with slight modification. The incubation mixture in a final volume of 2.0 ml contained: 0.8 μmole of albumin-bound [1-^{14}C]palmitate (the molar ratio of palmitate to albumin was 3:1) with a radioactivity of approximately 300,000 cpm; 80 μmole of phosphate buffer, pH 7.4; various concentrations of *Escherichia coli* 0127:B8 endotoxin (0–400 μg/ml); and freshly isolated myocytes representing 4–6 mg of protein. The incubation was carried out at 37°C in a metabolic shaker for 30 min. Metabolic $^{14}CO_2$ was collected by filter paper soaked in KOH, and the radioactivity was counted with 10 ml of 0.4% Omnifluor in toluene–ethanol (2:1, v/v) [1,2].

Oxidation of lactate (L form) or glucose by isolated myocytes was studied by incubating the freshly prepared myocytes with 15 μmole of [U-^{14}C]lactate (0.25 μCi) or 10 μmole of [U-^{14}C]glucose (0.3 μCi), respectively, in the presence of 80 μmole of phosphate buffer, pH 7.4, and various concentrations of endotoxin (0–400 μg/ml). The final volume of the incubation time was 30 min and the temperature was 37°C. The metabolic $^{14}CO_2$ was collected and counted as in the palmitate experiments.

The uptake of hexadecanol by isolated cardiac muscle cells was investigated according to the method of Spector and Soboroff [4]. The reaction mixture in a final volume of 2.0 ml contained 1.2 μmole of albumin-bound [1-^{14}C]hexadecanol (the molar ratio of hexadecanol to albumin was 1:47) with a radioactivity of approximately 80,000 cpm; 80 μmole of phosphate buffer, pH 7.4, various concentrations of *E. coli* 0127:B8 endotoxin (0–320 μg/ml), and freshly isolated cells containing 4–6 mg of protein. The incubations were carried out in 25-ml Erlenmeyer flasks for either 3 or 15 min at 37°C in a metabolic shaker. At the end of each incubation, the reaction mixture was quickly transferred to precooled (4°C) centrifuge tubes and was diluted with Krebs–Ringer's buffer (Ca^{2+}-free) to 14 ml. The tubes were then centrifuged at 60g for 4 min. The resulting myocyte pellets were washed twice using the same procedure and were suspended in PCS solution for radioactive counting.

Uptake of 3-O-methyl-D-glucose by isolated myocytes was estimated in a manner similar to that of hexadecanol. The incubation mixture in a final volume of 2.0 ml contained 40 μmole of 3-O-methyl-D[U-^{14}C]glucose (0.25 μCi), 80 μmole of phosphate buffer, pH 7.4, various concentrations of *E. coli* 0127:B8 endotoxin (0–320 μg/ml), and freshly prepared myocytes (4–6 mg of protein). The incubation time was 2 min and the temperature was 37°C.

E. coli 0127:B8 endotoxin (Boivin extraction) was obtained from Difco Laboratories, Detroit. [1-^{14}C]palmitic acid, [U-^{14}C]glucose, and Omnifluor were purchased from New England Nuclear Corporation. [1-^{14}C]cetyl alcohol (hexadecanol), 3-O-methyl-D[U-^{14}C]glucose, L[U-^{14}C]lactic acid, and PCS were products of Amersham/Searle Corporation. Collagenase (type 1 from *Clostridium histolyticum*), hyaluronidase (type 1-S from bovine testes), palmitic acid, L(+)-

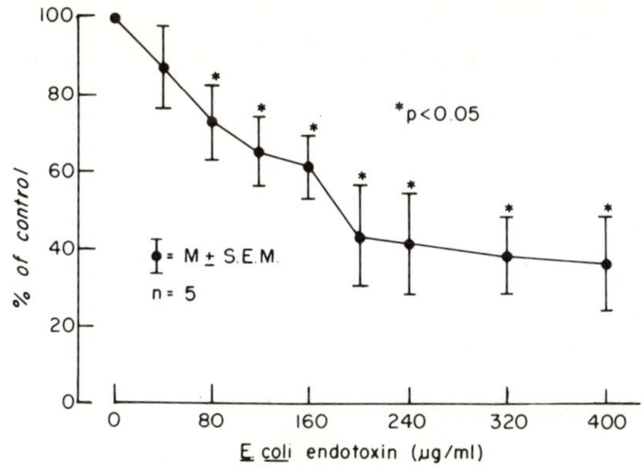

Fig. 1. In vitro effect of *E. coli* 0127:B8 endotoxin on palmitate oxidation by isolated adult cardiac muscle cells. The assay procedure is described under Materials and Methods. The ordinate represents percentage of control (zero endotoxin) of the rate of palmitate oxidation, and the abscissa indicates the amount of endotoxin present in the incubation mixture.

lactic acid, cetyl alcohol (hexadecanol), 3-O-methyl-D-glucose, and bovine albumin (fraction V, fatty acid free) were supplied by Sigma Chemical Company. D-Glucose and DL-carnitine were obtained from Fisher Scientific Company and Calbiochem, respectively. Other chemicals and reagents were of analytical grade.

A one-way analysis of variance was used for the statistical evaluation of the data. The protein concentration of isolated myocytes was measured according to Lowry et al [5] after the cell suspension was dissolved in 1 N NaOH solution.

RESULTS

Figure 1 shows the in vitro effect of *E. coli* 0127:B8 endotoxin on palmitate oxidation by isolated adult dog heart myocytes. Oxidation of palmitate by myocytes decreased with increasing concentrations of endotoxin; ie, the oxidation was decreased 26% ($P < 0.05$) by 80 μg/ml and 58% by 200 μg/ml of endotoxin. Unlike palmitate oxidation, lactate or glucose oxidation was increased by endotoxin as shown in Figures 2 and 3, respectively. Lactate oxidation was increased 31% ($P < 0.05$) by 160 μg/ml and 46% by 200 μg/ml of endotoxin, and the stimulation seemed to be saturable at higher concentration (Fig. 2). Glucose oxidation was increased proportionally with increasing concentrations of endotoxin; ie, a 32% ($P < 0.05$) and a 64% increase at 200 and 320 μg of endotoxin/ml, respectively. These results demonstrate that endotoxin in vitro affects myocardial substrate utilization by decreasing palmitate but increasing lactate and glucose oxidation.

Table I shows the effects of endotoxin on palmitate and lactate oxidation in the absence or presence of L-carnitine. It is apparent from Table I that the

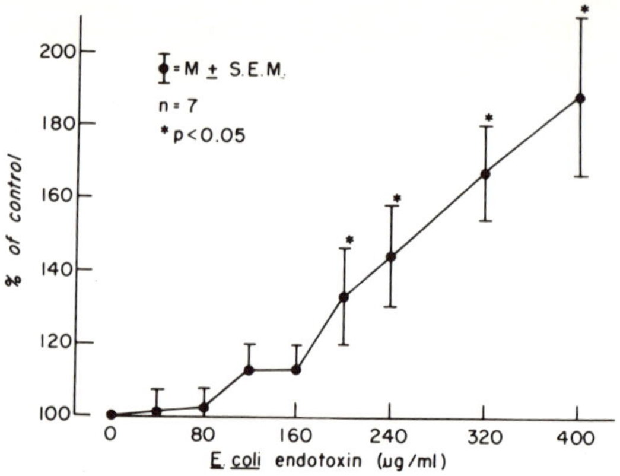

Fig. 2. In vitro effect of *E. coli* 0127:B8 endotoxin on lactate oxidation by isolated adult dog heart myocytes. For assay conditions, see Materials and Methods. The ordinate and the abscissa are the same as in Figure 1.

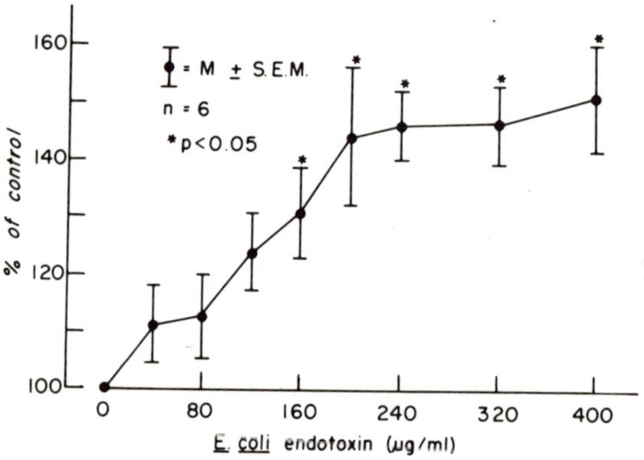

Fig. 3. In vitro effect of *E. coli* 0127:B8 endotoxin on glucose oxidation by isolated heart myocytes. The assay conditions are identical to those of Figure 1, except that glucose was used as a substrate.

oxidation of palmitate by isolated myocytes was decreased by 29% (162–115%/162%) in the presence of 4 mM L-carnitine. It should be pointed out that L-carnitine alone exhibited a stimulatory effect on palmitate oxidation. The increase in lactate oxidation induced by endotoxin was not affected by carnitine (nor did carnitine alone have any effect on lactate oxidation). These data suggest that endotoxin-induced changes in palmitate and lactate oxidation were not related to leakage of carnitine from myocyte in the course of the preparation.

TABLE I. Effects of Endotoxin on Palmitate and Lactate Oxidation by Isolated Myocytes in the Absence and Presence of L-Carnitine*

Addition	Palmitate oxidation (%)	Lactate oxidation (%)
None (control)	100	100
Endotoxin	69	130
Carnitine	162	106
Endotoxin and carnitine	115	129

*The experimental procedures were identical to those of Figures 1–3. Endotoxin, 200 μg/ml; carnitine, 4 mM. N = 4.

TABLE II. In Vitro Effect of *E. coli* Endotoxin on Hexadecanol Uptake and Palmitate Oxidation by Isolated Dog Heart Myocytes

Endotoxin concentrations (μg/ml)	Uptake of hexadecanol: 8-min incubation (%)	(N)	Uptake of hexadecanol: 15-min incubation (%)	(N)	Palmitate oxidation: 8-min incubation (%)	(N)
None (control)	100		100		100	(6)
20	97.7 ± 1.7	(4)	98.3 ± 2.5	(4)		
80	97.1 ± 1.8	(4)	97.6 ± 2.4	(4)		
200					48 ± 10*	(6)
240	95.4 ± 7.9	(4)	97.4 ± 6.2	(4)		
400	92.0 ± 3.4	(4)	94.5 ± 2.9	(4)		

*$P < 0.05$.

Table II shows the lack of effect of in vitro endotoxin on hexadecanol uptake by isolated dog heart myocytes. Endotoxin at the concentrations up to 400 μg/ml failed to affect hexadecanol transport, while palmitate oxidation was decreased by 52% ($P < 0.05$) during the same period of incubation time (8 min). Table III demonstrates that endotoxin was also ineffective in influencing 3-O-methyl-D-glucose uptake by isolated heart myocytes (up to concentrations of 320 μg/ml), while glucose oxidation was stimulated by 29% ($P < 0.05$) during the same incubation period (2 min).

DISCUSSION

It has been known for several years that the in vivo administration of *E. coli* endotoxin decreases myocardial oxidation of FFA and increases the utilization of lactate [6]. When heart homogenates were prepared from endotoxin-treated animals, lactate oxidation was again elevated, although palmitate oxidation was unchanged [1]. The in vitro addition of endotoxin to heart homogenates from

TABLE III. In Vitro Effect of *E. coli* Endotoxin on 3-O-Methyl-D-Glucose Uptake and Glucose Oxidation by Dog Heart Myocytes[a]

Endotoxin concentration (μg/ml)	Uptake of 3-O-methyl-D-glucose (%)	(N)	Glucose oxidation (%)	(N)
None (control)	100	(5)	100	(6)
40	109.6 ± 5.6	(5)		
80	102.1 ± 17.3	(7)		
160	107.7 ± 15.7	(5)		
200			129 ± 11[a]	(6)
240	100.8 ± 14.2	(5)		
320	103.2 ± 12.1	(5)		

[a]The incubation time for glucose experiments was 2 min.
[a]$P < 0.05$.

control dogs depressed the oxidation of both palmitate and lactate [2]. As isolated myocytes became available for the study of myocardial metabolism [3], it appeared that studies with these cells might shed further light on the action of endotoxin on myocardial metabolism.

Palmitate oxidation was decreased in a dose-dependent fashion in the presence of endotoxin. In an attempt to localize the site of this action, the effect of endotoxin on the uptake of the fatty acid analog, hexadecanol, was also studied. Since no effect was observed, it was concluded that transport across the plasma membrane is not involved in the effect of endotoxin on fatty acid oxidation. The addition of carnitine did not block the effect of endotoxin, indicating that the endotoxin-induced decrease in palmitate oxidation was not due to a loss of carnitine in the process of cell preparation. However, our data do not rule out the possibility of a defect in carnitine palmitoil transferase, a rate-limiting enzyme for acyl moiety transfer from cytosol to the mitochondria. Alterations in intramitochondrial oxidative processes may also be caused by endotoxin, as indicated by the findings that mitochondrial stage 3 respiration using palmitate as substrate was found to be decreased by this agent [7].

Although the site of action of endotoxin in influencing carbohydrate oxidation by the myocardium is not known, it seems that translocation of glucose into the cell is not involved, since the uptake of 3-O-methyl-D-glucose was not altered by endotoxin. Furthermore, the tricarboxylic acid cycle appears to be a likely site for the action as indicated by the fact that both glucose and lactate oxidation were similarly affected by endotoxin.

It is of interest to note that the in vivo effects of endotoxin on myocardial metabolism could be duplicated under in vitro conditions using isolated myocytes. Further investigations using this system may clarify the site and nature of endotoxin action on myocardial metabolism.

SUMMARY

The in vitro effects of *E. coli* 0127:B8 endotoxin on the oxidation of palmitate, glucose, and lactate by isolated, beating, adult dog heart myocytes were studied. The results were correlated with the in vitro effect of endotoxin on the transport of palmitate and glucose by using their respective analog, hexadecanol and 3-O-methyl-D-glucose. Endotoxin decreased palmitate but increased glucose and lactate oxidation in a dose-dependent manner. The changes in palmitate and lactate oxidation induced by endotoxin could not be corrected by addition of L-carnitine to the reaction mixture. The transport of hexadecanol or of 3-O-methyl-D-glucose was not affected by endotoxin. These results suggest that the endotoxin-induced alteration in palmitate and glucose oxidation by the heart is not the result of an impairment in metabolite transport, but rather a defect in the subsequent intracellular oxidative processes.

ACKNOWLEDGMENTS

This work was supported by the Office of Naval Research Contract No. N00014-76-C-0132 and National Institutes of Health grants HL 23618 (formerly HL 21223), AM 25270 (formerly AM 21052), and GM 26790.

REFERENCES

1. Liu MS, and Spitzer JJ: Myocardial fatty acid and lactate metabolism after *E. coli* endotoxin administration. Circ Shock 4:191, 1977.
2. Liu MS, Spitzer JJ: *In vitro* effects of *E. coli* endotoxin on fatty acid and lactate oxidation in canine myocardium. Circ Shock 4:181, 1977.
3. Liu MS, Spitzer JJ: Oxidation of palmitate and lactate by beating myocytes isolated from adult dog heart. J Mol Cell Cardiol 10:415, 1978.
4. Spector AA, Soboroff JM: Studies on the cellular mechanism of free fatty acid uptake using an analog, hexadecanol. J Lipid Res 13:790, 1972.
5. Lowry OH, Rosebrough NJ, Farr AL, and Randall RJ: Protein measurement with the Folin phenol reagent. J Biol Chem 193:265, 1951.
6. Spitzer JJ, Bechtel AA, Archer LT, Black MR, and Hinshaw LB: Myocardial substrate utilization in dogs following endotoxin administration. Am J Physiol 227:132, 1974.
7. Lavine L, Harano Y, DePalma RG: ATPase activity and palmitate oxidation of hepatic mitochondria in energy production in endotoxemia. Surg Forum 22:5, 1971.
8. Moss GS, Erve PO, Schumer W: Effect of endotoxin on mitochondrial respiration. Surg Forum 20:24, 1969.
9. DePalma RG, Harano Y, Robinson AV, Holden WD: Structure and function of hepatic mitochondria in hemorrhage and endotoxemia. Surg Forum 21:3, 1970.
10. Harris RA, Harris DL, Green DE: Effect of *Bordetella* endotoxin upon mitochondrial respiration and energized processes. Arch Biochem Biophys 128:219, 1968.
11. Mela L, Miller LD, Diaco J, Sugerman HJ: Effect of *E. coli* endotoxin on mitochondrial energy-linked functions. Surgery 68:541, 1970.

Metabolic and Endocrine Alterations Induced by Endotoxin and Sepsis at the Cellular Level

Judy A. Spitzer, Glenn J. Leach, and James A. Reeves

For several years we have been exploring modulation of cellular function by *Escherichia coli* endotoxin in isolated adipocytes. Use of the isolated adipocyte preparation affords one the opportunity to study mediated, as well as direct, effects of endotoxin. Experiments can be designed in two different modes: in vivo—ie, endotoxin is injected into an animal and after a suitable time interval, adipose tissue is removed and cells are isolated, or in vitro—when normal, untreated cells are exposed to endotoxin in vitro. In the latter case, one presumably measures only the results of a direct interaction between the cells and endotoxin.

By way of summary of this aspect of our previous work, alterations in some cell membrane-associated functions are presented in Table I.

Recently our interest has moved to the study of metabolic and hormonal alterations occurring in intraperitoneal sepsis. This shift in models stems from two considerations: (1) Even in this age of wide-spectrum antibiotics, intraperitoneal sepsis is a life-threatening clinical syndrome. Intraperitoneal sepsis plays an important role even in the adult respiratory distress syndrome, where in 55% of the cases a remote site of infection, most commonly localized in the intraperitoneal space, is responsible for the observed pulmonary failure [8]. (2) Septic shock may be an exception to the general cardiovascular shock rule in that during high flow states in septic shock, normal or even higher than normal cardiac output may exist. Under such conditions, the conventionally employed therapeutic measures aimed at correcting deteriorating cardiopulmonary function do not accomplish the desired results, and the irreversibility of the shock is thought to be due to cellular metabolic failure. Thus, septic shock represents a special challenge to investigators interested in aspects of metabolic and hormonal regulation at the cellular level. In this context then, it becomes important that we

Pathophysiological Effects of Endotoxins at the Cellular Level, pages 123–130

TABLE I. Modulation of Cellular Function by *E. coli* 0127:B8* Endotoxin in Isolated Adipocytes

Function	Description of change	Produced in vivo	Produced in vitro
Norepinephrine-stimulated lipolysis	Increase	(1)[a]	(2)
cAMP accumulation	Increase (in the absence of any hormonal stimulation)		(2)
Ca^{2+} fluxes	Some similarities to action of ionophore A23187		(3)
Insulin sensitivity	Decrease (in terms of glucose oxidation) Increase (in terms of antilipolysis)	(4,5)	(5,6)
Insulin binding	Increased affinity at low insulin concentrations	(5)	
3-O-Methylglucose transport	Increase	(7)	

*Difco Boivin extraction.

[a]Numbers in parentheses refer to the appropriate references.

compare our endotoxin results with the cellular changes occurring in an experimental model of intraperitoneal sepsis closely simulating clinical peritonitis, a condition which includes an endotoxin component. In this report, we present data on the response to norepinephrine and insulin of adipocytes taken from rats in the early phase of sepsis and compare these responses to those found in experimental endotoxin shock.

METHODS

Sepsis was induced by cecal ligation and perforation under ether anesthesia in male Sprague–Dawley rats [9]. Blood sampling, isolation of adipocytes from epididymal fat pads, and sacrificing the fasted animals followed 12–15 hr later. The same schedule applied to sham-operated controls. Nonoperated controls were fasted for 12–15 hr. Intraperitoneal sepsis, caused by infection with both Gram-negative and Gram-positive organisms was documented by cultures of blood and peritoneal samples. Plasma glucose, insulin, and FFA concentrations were determined [5,10], and glucose oxidation as well as the lipolytic response of isolated fat cells were measured using procedures described previously [6]. Some metabolic data pertaining to this septic rat model were presented in a preliminary report [11].

TABLE II. Plasma Glucose, Insulin, and FFA Concentrations

Treatment	Glucose (mg/dl)	Insulin (IRI) (μU/ml)	FFA (mEq/liter)
Septic	124.5 ± 3.4 (28)[a]	22.4 ± 3.18** (14)	0.529 ± 0.039 (10)
Sham operated	117.6 ± 3.7 (24)	12.8 ± 1.96* (16)	0.476 ± 0.042 (8)
Fasted	59.3 ± 3.8*** (14)	7.8 ± 1.0 (4)	—

[a]Numbers in parentheses refer to the number of animals.
*$P < 0.025$ vs fasted.
**$P < 0.0005$ vs fasted and $P < 0.01$ vs sham.
***$P < 0.0005$ vs septic and sham.

RESULTS AND DISCUSSION

Plasma Studies

Glucose and FFA concentrations of septic and sham-operated animals were not different from one another. Plasma insulin levels, however, were significantly elevated in the septic as compared with the sham-operated rats (Table II). Significantly more circulating insulin was required in the septic rats to maintain a blood glucose level essentially unchanged from that of the sham-operated animals. Such an occurrence is compatible with the development of total body insulin resistance.

Studies with Isolated Adipocytes

Glucose oxidation. There was no difference among the three experimental groups in the basal rate of oxidation of [^{14}C]glucose (Table III). Insulin-stimulated glucose oxidation, however, was significantly depressed in cells of septic rats at all concentrations of the hormone used, thus indicating insulin resistance of these adipocytes.

Lipolytic response. The lipolysis-enhancing effect of norepinephrine as well as the lipolysis-suppressing, or antilipolytic, effect of insulin were both studied. Results of these experiments are summarized in Table IV. Basal lipolysis by septic adipocytes was significantly higher than that of sham-operated or fasted, nonoperated control-derived cells. The extent of norepinephrine-stimulated lipolysis was unchanged in the three groups. Sensitivity to norepinephrine, assessed by the ratio of norepinephrine-stimulated/basal lipolysis, however, was significantly reduced in septic versus fasted control cells, in terms of both FFA

TABLE III. Glucose Oxidation

Treatment	[^{14}C]Glucose oxidized (nmol/hr/2 × 10^5 cells)				
Insulin (ng/ml)	0	0.125	0.250	1.25	5.00
Septic	1.972 ± 0.426 (14)[a]	3.562 ± 1.010* (7)	3.528 ± 0.247** (7)	5.794 ± 0.651*** (14)	6.465 ± 0.659*** (14)
Sham-operated	3.515 ± 0.911 (15)	7.533 ± 1.397 (6)	7.075 ± 1.249 (8)	11.067 ± 1.014 (14)	11.688 ± 1.096 (14)
Fasted	1.962 ± 0.487 (4)	—	9.653 ± 1.811 (5)	15.242 ± 2.423 (5)	15.628 ± 2.916 (5)

[a]Numbers in parentheses refer to the number of animals.
*P < 0.025.
**P < 0.010 vs sham; P < 0.005 vs fasted.
***P < 0.0005.

TABLE IV. Lipolytic Response (μmoles released/hr/10^6 cells)

Addition to medium[a]		Septic		Sham		Fasted	
NE	Insulin	FFA	G	FFA	G	FFA	G
–	–	0.568 ± 0.214 (10)[b]	0.324 ± 0.089* (11)	0.257 ± 0.062 (10)	0.139 ± 0.021 (9)	0.158 ± 0.033 (4)	0.091 ± 0.023 (3)
+	–	3.237 ± 0.835 (11)	1.724 ± 0.445 (11)	2.681 ± 0.557 (13)	1.629 ± 0.324 (12)	4.763 ± 0.980 (6)	2.391 ± 0.254 (6)
–	+	0.574 ± 0.149** (11)	0.162 ± 0.025 (9)	0.132 ± 0.019*** (8)	0.105 ± 0.017 (9)	—	—
+	+	1.323 ± 0.339 (11)	0.405 ± 0.068 (11)	0.821 ± 0.137 (12)	0.326 ± 0.042 (12)	1.443 ± 0.434 (5)	0.402 ± 0.104 (5)

[a]NE, norepinephrine, 0.1 μg/ml; and insulin, 125 μU/ml.
[b]Numbers in parentheses refer to the number of animals.
*$P < 0.05$ vs sham and $P < 0.025$ vs fasted.
**$P < 0.005$ vs sham.
***$P < 0.05$ vs basal.

TABLE V. Norepinephrine (NE) Stimulation of Lipolysis in Adipocytes of Septic, Sham-Operated, and Fasted Rats

Treatment	NE/basal lipolysis G	NE/basal lipolysis FFA
Septic	5.19 ± 1.30*** (11)[a]	4.02 ± 0.700* (8)
Sham	9.72 ± 4.80** (8)	9.84 ± 2.64 (8)
Fasted	22.46 ± 1.65 (3)	36.20 ± 14.65 (4)

[a]Numbers in parentheses refer to the number of animals.
*$P < 0.05$ vs sham and fasted.
**$P < 0.025$ vs fasted.
***$P < 0.0005$ vs fasted.

and glycerol release, and in septic versus cells of sham-operated control rats in terms of FFA release (Table V).

Insulin caused similar depressions of the norepinephrine-stimulated lipolytic response in cells of septic as well as sham-operated rats (Table IV). On the other hand, insulin produced nonuniform antilipolytic effects on basal lipolysis (Table VI). Glycerol release was significantly reduced in septic cells, while FFA release remained unaltered or elevated. One possible explanation of this phenomenon would be accentuated, partial lipolysis in the septic cells. Conversely, in fat cells of sham-operated control rats, insulin significantly depressed basal FFA release, while glycerol release was unchanged. The latter finding is consonant with the possibility of increased reesterification.

The isolated adipocyte model proved to be a useful tool in studying alterations in metabolic fuel utilization and hormone sensitivities in experimentally produced sepsis. Lowered insulin sensitivity of these cells in terms of glucose oxidation reflected the observed total body insulin resistance. Under basal conditions, lipid mobilization in septic adipocytes proceeded at a significantly higher rate than in cells of sham-operated and nonoperated control rats. Furthermore, the cells maintained this elevated FFA release even in the presence of insulin. Septic adipocytes also appeared to have lost a part of their norepinephrine-stimulated lipolytic response, as documented by the significantly reduced ratio of norepinephrine-stimulated/basal lipolysis.

In summary, the results indicate (1) insulin resistance in the whole animal as well as at the adipocyte level in terms of glucose oxidation and basal FFA release, and (2) altered patterns in the lipolytic–antilipolytic responses of isolated adipocytes of 12- to 15-hr septic rats.

TABLE VI. Suppression of Basal FFA and Glycerol (G) Release by Insulin (μmoles released/hr/10^6 cells)

	Septic				Sham	
	FFA	G	FAA/G	FFA	G	FFA/G
Basal	0.568 ± 0.214 (10)[a]	0.324 ± 0.089 (11)	2.71 ± 0.62 (7)	0.257 ± 0.062 (10)	0.139 ± 0.019 (8)	2.20 ± 0.60 (7)
+ Insulin	0.574 ± 0.149 (11)	0.162 ± 0.025* (9)	3.68 ± 1.04 (8)	0.132 ± 0.019* (8)	0.105 ± 0.017 (9)	1.23 ± 0.019* (7)

[a]Numbers in parentheses refer to the number of animals.
*$P < 0.05$ vs basal.

Comparing some of the cellular metabolic consequences of endotoxin administration with those of experimental peritonitis, differences as well as some similarities are evident. In endotoxic shock, norepinephrine-stimulated lipolysis is accentuated, but so is the repression of this lipolysis by insulin. In septic shock, neither of these phenomena is observed. Basal lipolysis in septic adipocytes is enhanced; in endotoxic fat cells it is not. In terms of insulin action in adipocytes, reduced responsiveness of glucose oxidation to the stimulation by this hormone is a feature shared in common by both shock models. Since the microbial flora in the rat experimental peritonitis model contains a significant proportion of Gram-negative, endotoxin-producing organisms, it is reasonable to assume that some of the pathological changes observed at the cellular level can be attributed to the action of endotoxins.

ACKNOWLEDGMENTS

This work was supported by ONR Contract N00014-76-C-0133. Appreciation is expressed for the skillful technical assistance rendered by Mr. P. J. O'Malley. Mr. J. A. Reeves was the recipient of a Summer Medical Student Fellowship of the American Heart Association/Louisiana, Inc.

REFERENCES

1. Spitzer JA, Kovach AGB, Rosell S, Sandor P, Spitzer JJ, Storck R: Influence of endotoxin on adipose tissue metabolism. In Kovach AGB, Stoner HB, Spitzer JJ (eds): "Neurohumoral and Metabolic Aspects of Injury." New York: Plenum Press, 1973, p. 337.
2. Spitzer JA: Endotoxin-induced alterations in isolated fat cells: Effect on norepinephrine-stimulated lipolysis and cyclic 3′, 5′-adenosine monophosphate accumulation. Proc Soc Exp Biol Med 145:186, 1974.
3. Hikawyj-Yevich I, Spitzer JA: The role of adrenergic receptors and Ca^{2+} in the action of endotoxin on human fat cells. J Surg Res 23:233, 1977.
4. Spitzer JA: Altered insulin sensitivity in endotoxemia. Fed Proc 38:1193, 1979.
5. Holley DC, Spitzer JA: Insulin action and binding kinetics in adipocytes exposed to endotoxin in vitro and in vivo. Circ Shock 7:3, 1980.
6. Spitzer JA, Holley DC: Alterations in insulin action by endotoxin in vitro. In Schumer W, Spitzer JJ, Marshall BE (eds): "Advances in Shock Research." Vol 2. New York: Alan R. Liss, 1979, p. 129.
7. Leach GJ, O'Malley P, Spitzer JA: Endotoxin induced alterations in glucose transport in rat adipocytes. Circ Shock 6:168, 1979.
8. Richardson DJ, Frye DE, Van Arsdall L, Flint LM: Delayed pulmonary clearance of Gram-negative bacteria: The role of intraperitoneal sepsis. J Surg Res 26:499, 1979.
9. Chaudry IH, Wichterman KA, Baue AE: Effect of sepsis on tissue adenine nucleotide levels. Surgery 85:205, 1979.
10. Itaya K, Ui M: Colorimetric determination of free fatty acids in biological fluids. J Lipid Res 6:16, 1965.
11. Spitzer JA, Leach GJ, Reeves JA: Altered insulin and catecholamine sensitivities in adipocytes of septic rats. Presented at the Meeting of the Southwest Section, Soc Exp Biol Med, Oct 26–27, 1979, Houston, Texas.

Functional Alterations in Endotoxemia

Interactions of Bacterial Lipopolysaccharides With Tissue Macrophages and Plasma Lipoproteins

Ronald V. Maier, John C. Mathison, and Richard J. Ulevitch

Sepsis continues to be a major cause of morbidity and mortality in the United States. Mortality due to infectious etiologies was responsible for more than 125,000 deaths in 1976 and represented the fourth leading cause of death overall. Greater than 85% of these deaths are attributable to bacterial and mycobacterial infections with greater than 66% of the bacterial sources being due to Gram-negative bacillae. In the face of marked advances in medical therapy, septicemia-related mortalities continue to increase at about 5% per year, with a persistent approximate 50% mortality associated with a polymicrobial septicemic state [1].

The endotoxin complex of lipid, polysaccharide, and small amount of associated protein has been shown to be a ubiquitous integral part of the Gram-negative bacterial cell wall and appears to be released only upon lysis of the bacterium [2, 3]. Endotoxin and, more specifically, lipopolysaccharides (LPS), with the lipid A-associated protein (LAP) having been removed, have been shown to reproduce in the experimental host the spectrum of pathophysiological changes which closely resemble those seen clinically in the septic patient (reviewed in [4]). For example, we have shown that the intravenous bolus injection of 250 μg of *Salmonella minnesota* R595 or *Escherichia coli* 0111:B4 LPS in 2.5-kg rabbits produces acute neutropenia, thrombocytopenia, complement depletion, hypotension, DIC, and death within 12 hr [5].

The purpose of the present work was to explore several aspects of the host–LPS interaction both intravascularly and at the tissue level. The development of a method for covalently binding radioiodine to LPS without altering its biophysical or biologic properties [6] has permitted a more exact evaluation of the clearance,

Pathophysiological Effects of Endotoxins at the Cellular Level, pages 133–155

tissue distribution, and cellular and subcellular localization of intravenously injected LPS. The initial rapid tissue distribution of LPS is to the lung, liver, and spleen. The use of autoradiography demonstrates that the major cellular target of LPS is the fixed tissue macrophage (Mφ) of liver and spleen and has been shown to be concentrated within the phagocytic vacuoles [7]. To evaluate the role of the LPS–hepatic macrophage (H–Mφ) interactions in the pathogenesis of shock and DIC, a technique has been developed for the isolation and maintenance in culture of H–Mφ. Subsequently, the treatment of explanted H–Mφ with LPS has demonstrated a selective response with the increased production of several cellular enzymes. Also, both in vivo and in vitro treatment with LPS has led to evidence for a direct H–Mφ toxicity with decreased yields and increased cell loss from cultures [8]. Finally, the LPS remaining in plasma after the initial rapid clearance was found to be complexed with high-density lipoprotein (HDL) and demonstrated a prolonged half-life, decreased density, and inhibition of several endotoxic properties when compared to the parent LPS [9]. The potential effects on the host of these interactions will be discussed in relation to the initiation of the basic pathophysiology of endotoxemia.

PROCEDURES AND RESULTS

Lipopolysaccharide

For all experiments either the smooth, *E. coli* 0111:B4 or the rough mutant, *Salmonella minnesota* R595 LPS were used. The LPS was prepared in our laboratories using the hot phenol–water extraction (Westphal) for 0111:B4 and the phenol–chloroform–petroleum ether method of Galanos et al [10] for R595 LPS. These highly purified LPS preparations are free of LAP but contain a high percentage of the biologically active lipid A. The LPS was covalently radioiodinated using the method of Ulevitch [6] and was found to remain bound to the LPS for the duration of the experiments [7].

Clearance of LPS

The clearance of LPS from the intravascular compartment was followed in rabbits using 250 μg of either ^{125}I-0111:B4 or ^{125}I-R595 LPS given as an intravenous bolus in 1 cc of normal saline (NS). Using the expected blood volume for 2.5- to 3.0-kg rabbits, the initial level of LPS should be 1.5–2.0 μg/ml. As shown in Figure 1, the clearance of both 0111:B4 and R595 LPS from the blood is rapid with only 60% of 0111:B4 and 30% of R595 doses remaining at 1 min. After the initial rapid clearance, the LPS is cleared much more slowly with a half-life of $>$ 10 hr. Greater than 95% of the slowly removed ^{125}I-LPS was found in cell-free plasma and demonstrated electrophoretic mobilities identical to that of the parent LPS in a sodium dodecyl sulfate–polyacrylamide gel electrophoresis (SDS–PAGE) system [11].

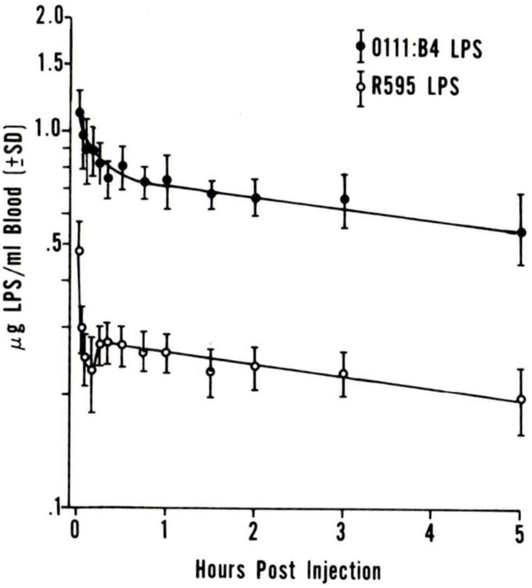

Fig. 1. Clearance of either 0111:B4 or R595 LPS from the blood of rabbits after iv injection of 250 μg of ^{125}I-LPS. The estimated initial LPS concentration based on blood volumes would be 1.5–2.0 μg/ml. Each point represents data from six animals.

Tissue Distribution of LPS

In Figures 2 and 3, the tissue distribution for 0111:B4 and R595 ^{125}I-LPS are shown. A 250-μg dose of LPS containing 10 μCi of ^{125}I-LPS was used in 2.5-kg rabbits, which were sacrificed at 5 or 180 min post injection. After 5 min, 55% of the 0111:B4 and 17% of the R595 LPS remained in the blood. The tissue distributions were very similar with liver localization responsible for the majority of uptake of about 40% for 0111:B4 and of about 50% for R595 LPS. This pattern persisted at 180 min with slight decreases in concentration in most tissues assayed, with 33% of 0111:B4 and 10% of the R595 LPS persisting in the blood. Overall, greater than 80% of all radioactivity injected could be recovered in the tissues shown.

The concentration of ^{125}I-LPS in each of the tissues is displayed in Figures 4 and 5, and again the patterns for 0111:B4 and R595 LPS are similar. The highest concentrations of 0111:B4 LPS were found in liver and spleen, with the only significant change from the 5- to 180-min sampling being a four-fold increase in the concentration of LPS in the adrenal. R595 LPS showed a similar high concentration of ^{125}I-LPS in both liver and spleen but, unlike 0111:B4, there was also a significant concentration in the lung. Also, at the 180-min time point, there was no increase in adrenal concentration of R595, and the concentration in lung tissue had decreased significantly. The bile, by 180 min, contained

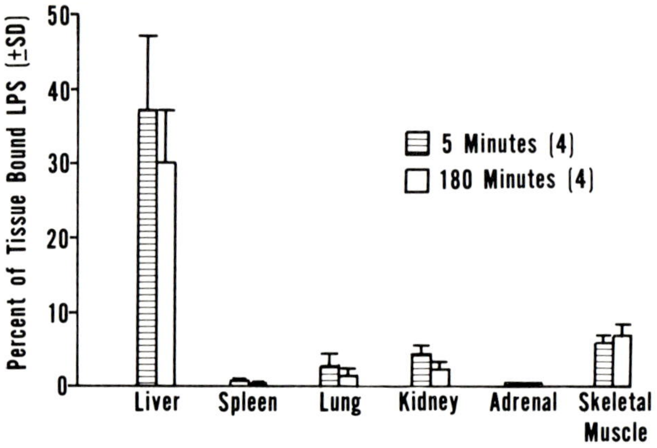

Fig. 2. Distribution of tissue-bound 0111:B4 LPS at 5 and 180 min post iv injection of 250 μg of ^{125}I-LPS. Number of animals is in parentheses.

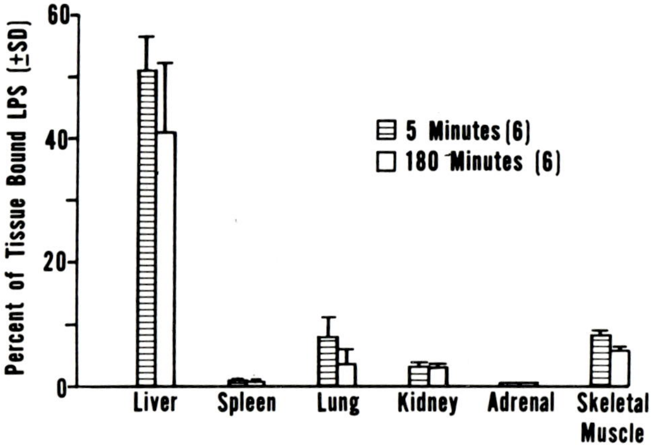

Fig. 3. Distribution of tissue-bound R595 LPS at 5 and 180 min post iv injection of 250 μg of ^{125}I-LPS. Number of animals is shown in parentheses.

concentrations of both R595 and 0111:B4 LPS equal to that seen in the liver. Injection of sublethal doses of either LPS produced similar tissue distribution patterns.

Cellular Localization of LPS

Tissue samples from liver, spleen, lung, kidney, adrenal, cardiac, and skeletal muscle, lymph node, and small intestine were examined using autoradiography on both the light-microscopic and electron-microscopic levels [7, 12, 13]. The

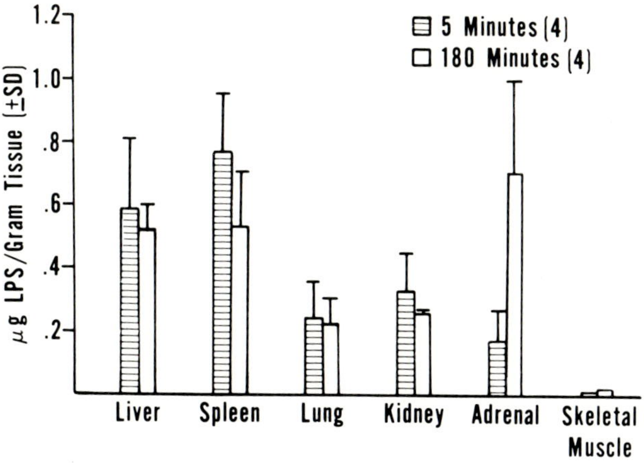

Fig. 4. Concentration of tissue-bound 0111:B4 LPS at 5 and 180 min post iv injection of 250 μg of ^{125}I-LPS in 2.5-kg rabbits.

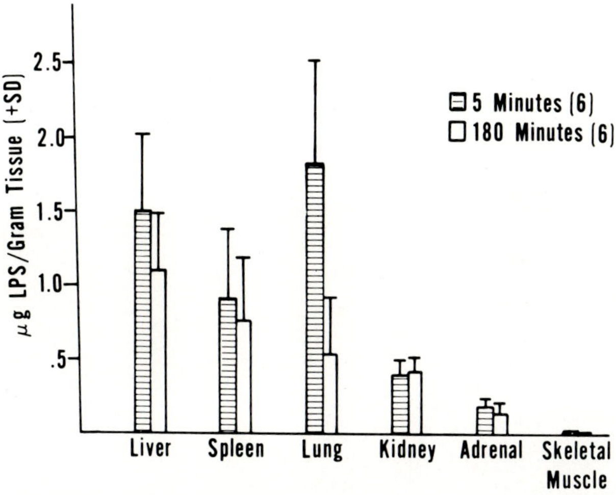

Fig. 5. Concentration of tissue-bound R595 LPS at 5 and 180 min post iv injection of 250 μg of ^{125}I-LPS in 2.5-kg rabbits. The only significant change in concentration was noted in lung tissue.

rabbits had been treated 5 or 180 min prior to sacrifice with either 0111:B4 or R595 LPS as described above. Similar cellular patterns were noted for both LPS preparations, except where noted. The liver showed localization in the H–Mϕ, while hepatocytes and endothelial cells were negative. Electron microscopy of the 180-min samples showed concentration of the LPS within the phagocytic vacuoles of the H–Mϕ (Fig. 6). Splenic sinusoids demonstrated diffuse labeling

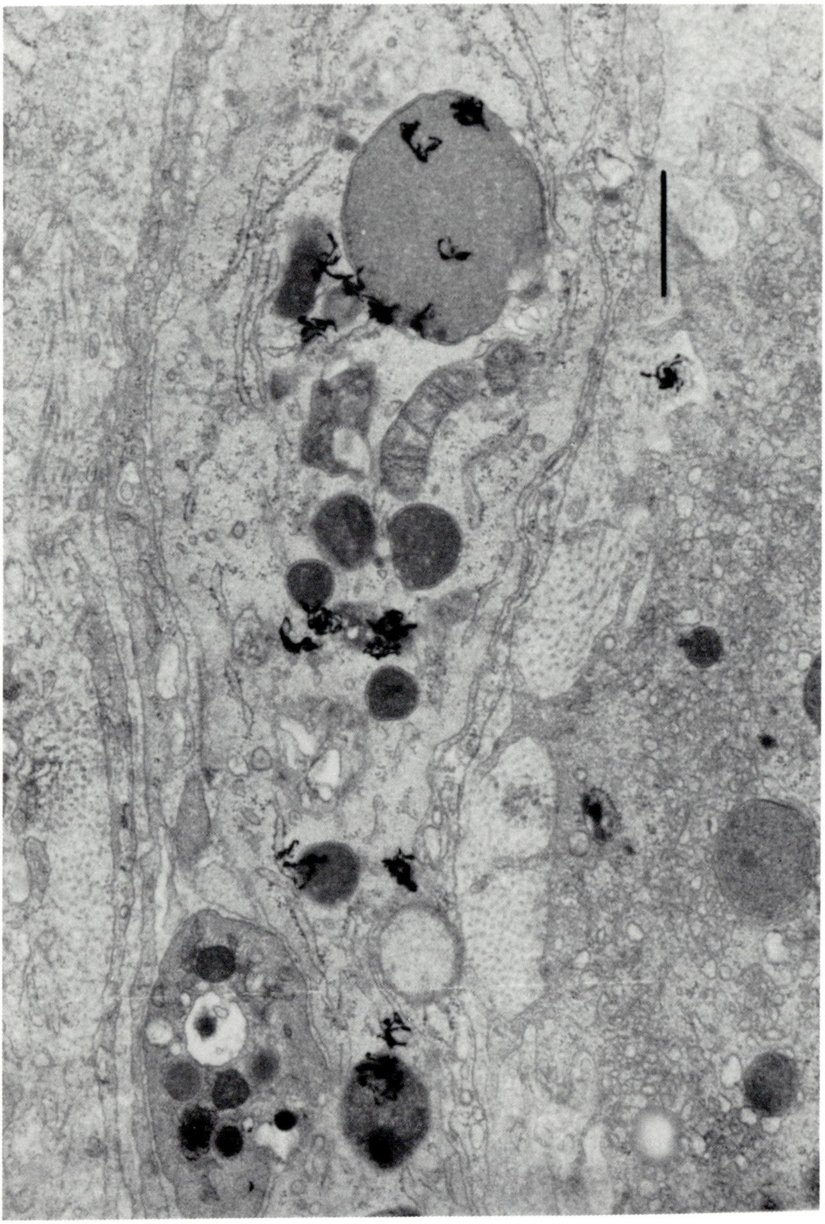

Figure 6

with LPS which was localized to neutrophils and Mϕ contained within the sinusoids. Rare labeling was noted over lymphoid follicles at any time.

The pulmonary capillaries contained large numbers of neutrophils and monocytes at 5 min post treatment with R595 LPS, and a small percentage of cells were heavily labeled. Both total number and number of labeled cells decreased by 180 min post R595 treatment. Treatment with 0111:B4 LPS produced a much smaller pulmonary intravascular accumulation of cells and few cells were observed to contain ^{125}I-LPS. Examination of adrenal tissue revealed that label in R595 LPS-treatment animals was associated with the sinusoidal Mϕ. In contrast, 0111:B4 LPS treatment produces a marked increase in label over the cytoplasm of adrenal cortical cells at the 180-min time point. The remaining tissues did not show specific cellular labeling, except for an occasional intravascular phagocytic cell.

Recovery of LPS from Liver

Homogenates of liver samples from animals treated 180 min prior with 0111:B4 or R595 LPS were coelectrophoresed with the parent LPS, which was labeled with ^{131}I. Using the SDS–PAGE system [11], identical electrophoretic patterns were obtained for the respective parent and liver homogenate LPS. Greater than 80% of the radiolabel within the liver sample was recovered and the greater than 80% recovery from gel slices confirms that the ^{125}I had not dissociated in vivo from the LPS. Also, there was no evidence of degradation of the LPS moiety after 3 hr of localization within the H–Mϕ [7].

Isolation of H–Mϕ

The technique developed for isolation of rabbit H–Mϕ is a modification of the work of Mills and Zucker-Franklin [14] and others [15]. Perfusion of the liver in situ with saline removes the vast majority of intravascular cellular elements, and distention with 200 cc of a 0.01% collagenase solution produces grossly visible disruption of the hepatic structure. The subsequent digestion of samples of liver tissue in 0.2% Pronase–Hanks' balanced salt solution (HBSS) produces a selective destruction of the hepatocytes as shown by greater than 95% of the hepatocytes being lysed or positive for trypan blue uptake at the completion of a 90-min digestion. The heterogeneous population of cells (nonparenchymal cells, NPC) obtained after multiple washings of the Pronase digest, in contrast, is greater than 95% viable by trypan blue exclusion analysis. The NPC consists of about 30% Mϕ, 50% endothelial cells, and 20% lymphocytes, smooth muscle

Fig. 6. Electronmicroscopic autoradiograph of a H–Mϕ obtained from a rabbit 180 min post iv injection of 250 μg of ^{125}I-R595 LPS. Similar results were obtained using ^{125}I-0111:B4 LPS. The silver grains are associated primarily with phagocytic vacuoles. Marker, 1μm.

cells, fibroblasts and occasional neutrophils [16]. The H–Mϕ was further enriched by its enhanced adherence properties to glass in overnight culture. This produced an adherent cell population with about 80–85% Mϕ. The resistance to trypsin treatment of glass-adherent Mϕ allowed for a final treatment of the cultures with Enzar-T (a trypsin–chymotrypsin solution) to remove the non-Mϕ-contaminating cells and leave a population of cells in culture of greater than 95% Mϕ. This final population of cells could be maintained in culture for up to 10 days without outgrowth of other cell types. The technique employed routinely produces about 5×10^6 NPC/g and, after final enzymatic treatment, produces about 6×10^5 adherent cells/g of wet-weight liver in culture.

Characterization of H–Mϕ

The final adherent cell population was characterized to ensure that a pure population of H–Mϕ was present. Using phase microscopy, two distinct morphologic cell populations were found: One was smaller, round, and less mobile, while the other was larger, fusiform, and more mobile. The cytoplasm contained numerous inclusion bodies, which increase in number with time in culture, and an ovoid indented, eccentrically placed nucleus (refer Fig. 7). There was no evidence of mitotic activity detected using phase microscopy, total DNA content of the cultures, or [^{3}H]-thymidine uptake (Maier, unpublished data).

Using positive reactivity in fluorescein diacetate assays [17] and total DNA content [18] of parallel cultures, the H–Mϕ population viability was greater than 90% on Day 1 in culture and decreased to approximately 80–85% by Day 4. The intense diffuse cytoplasmic nonspecific esterase activity indicative of other Mϕ populations [19] was found in greater than 95% of the H–Mϕ culture cells and remained unchanged for up to 10 days. Phagocytic capacity was tested using both untreated latex beads and opsonized zymosan. Both particles were avidly ingested by greater than 95% of the adherent cells. Finally, Mϕ are known to possess both Fc and C3b membrane receptors [20]. The H–Mϕ were greater than 90% positive for Fc receptors after 24 hr and for C3b receptors after 72 hr in culture, as demonstrated by the ability of the cell to rosette appropriate ligand-bearing erythrocytes [8].

Response of H–Mϕ to LPS

The response of explanted H–Mϕ to LPS was examined using three experimental groups: untreated, latex bead-treated, and cultures exposed to graded doses of LPS. Although the data reported were obtained using the lipid A-rich *Salmonella minnesota* R595 LPS, similar results were found using *E. coli* 0111:B4 LPS. LPS, at all concentrations used, produced an increase in size, perinuclear granularity, and peripheral vacuolation with a more rounded cellular

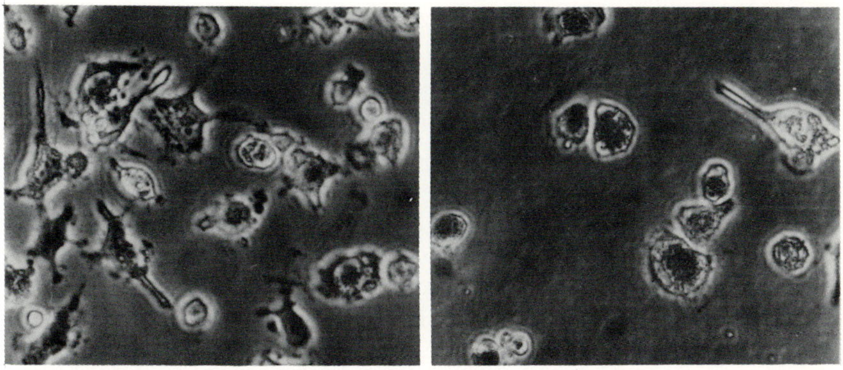

Fig. 7. Phase-contrast micrographs of rabbit H–Mφ cultures. Left panel is typical of baseline untreated cultures after 48 hr. The adherent cells are characteristic of other Mφ populations. Right panel represents H–Mφ in culture 24 hr after addition of R595 LPS (10 μg/ml) with peripheral vacuolation and increased granularity demonstrated.

TABLE I. H–Mφ Viability under Varying Culture Conditions

	Fluorescein assay		DNA level	
Culture conditions	20 hr	96 hr	20 hr	96 hr
Untreated	93[a]	85	98 ± 6[b]	85 ± 4
Latex beads	93	89	96 ± 6	87 ± 6
R595 LPS (μg/ml)				
1	84	65	93 ± 5	69 ± 1
10	78	65	89 ± 5	67 ± 7
100	77	60	79 ± 7	60 ± 8

[a]Numbers are percentage of cells viable, as determined by the retention of fluorescein fluorescence after hydrolysis of fluorescein diacetate by H–Mφ, and are based on at least five cell counts of 100 cells each which varied by less than 10% at each time point.

[b]Numbers are percentage of DNA remaining normalized to initial values being 100% and demonstrate the loss of adherent H–Mφ in parallel cultures. Each point represents the mean ± SEM for at least six experiments.

morphology (Fig. 7). Neither the percentage nor the avidity of the Mφ capable of latex bead phagocytosis was altered by LPS treatment.

Cell viability, as determined by a fluorescein diacetate assay [17], for the varying culture conditions is summarized in Table I. Exposure to LPS produces a 30–40% decrease in viability by Day 4 in contrast to a 10–15% decrease in untreated and latex bead-treated cultures. Also, as shown in Table I, there is a concomitant, parallel increased loss of DNA from cultures treated with LPS. Significant loss of DNA is noted with LPS doses as low as 1 μg/ml.

To determine if LPS treatment in vitro stimulated the explanted H–Mϕ, we studied both the total activity as well as the percentage of release of several different types of cellular enzymes. The acid hydrolases, β-glucuronidase and acid phosphatase; the cytoplasmic marker, LDH; the constituitive Mϕ enzyme, lysozyme; and a neutral proteinase, plasminogen activator (PA) were evaluated. The response of the acid hydrolases is depicted in Figure 8 for both untreated cultures and LPS-treated cells (R595, 10μg/ml). For these data and the data to follow, treatment of cultures with latex beads did not alter enzyme levels from untreated levels. Thus, for clarity, only untreated culture values will be presented. Both hydrolases demonstrate two to three fold increase in activity in control cultures over the 4-day period. LPS treatment slightly increased the total amount of acid phosphatase, while decreasing the level of β-glucuronidase when compared to control cultures.

In marked contrast to the above data, the addition of as little as 1 μg/ml of LPS produced a five to six fold increase in the total activities of LDH, lysozyme, and PA in the H–Mϕ cultures over the experimental period. The control cultures showed a 1.5 to two fold increase in total activity similar to that seen for the acid hydrolases (refer to Fig. 9).

The possibility that these differences in enzyme levels were due to differential inactivation of secreted enzyme in the culture medium was examined using cell-free culture medium supernatants. LDH, β-glucuronidase, and lysozyme were all stable for at least 96 hr at 37°C, with each showing less than 10% variation in activity. PA showed a significant loss in activity with about 60% of initial activity remaining by Day 4 but, again, demonstrated no differential loss of activity in control versus LPS-treated cultures. However, a corrected total activity for PA would be even higher than that shown in Figure 9. Finally, acid phosphatase demonstrated not only a loss in activity (about 65–85% of activity remaining on Day 4), but also a differential inactivation with activity in LPS supernatants showing about a 20% greater loss than control supernatants. Using this factor for correction, the total activity in LPS-treated cultures would exceed control levels by 70% on Day 4.

The percentage of total activity released into the medium was measured for each enzyme to evaluate any selective increased release by H–Mϕ in response to LPS treatment. For PA and lysozyme, greater than 85% of activity was released by 72 hr in culture, as has been shown by others [21, 22]. This release was unaffected by phagocytosis of latex or treatment with the entire dose range of LPS. The percentage of release of LDH, acid phosphatase, and β-glucuronidase is shown in Figure 10. Each enzyme demonstrates a similar 20–25% increased percentage of release in LPS-treated cultures, with latex ingestion again not altering percentage of release from that seen in control cultures. The increased release of LDH, a marker for cellular toxicity, correlates well with the 20–25% increased loss of DNA from LPS-treated cultures.

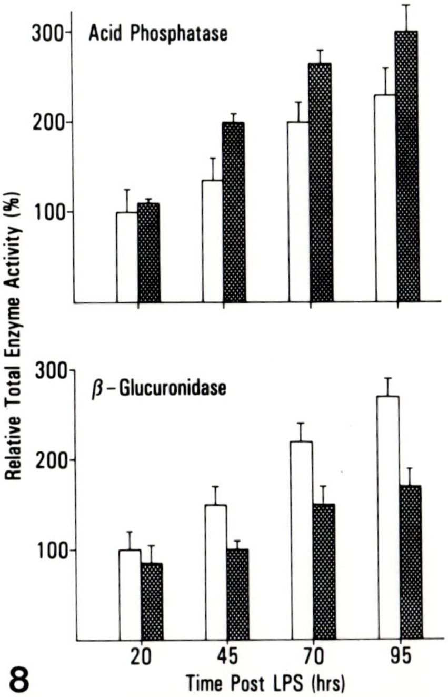

8

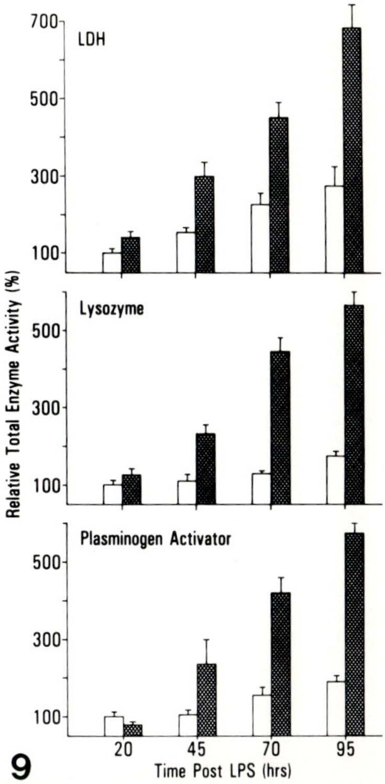

9

Fig. 8. Response in levels of acid phosphatase and β-glucuronidase in explanted H–Mϕ treated with R595 LPS (10 μg/ml) at Time 0. The data for untreated (open bars) and LPS-treated (hatched bars) cultures represent the mean + 1 SEM for at least five separate rabbit preparations and is based on initial control levels being 100%

Fig. 9. LDH, lysozyme, and PA responses in total enzyme activity in explanted H–Mϕ treated with R595 LPS (10 μg/ml at Time 0. The relative activity at each time point is related to initial control culture levels being 100% and each represents the mean + 1 SEM for five separate experiments. Untreated culture values are represented by open bars and LPS-treated cultures by hatched bars.

Interaction of LPS with Plasma

The addition of either *E. coli* 0111:B4 or *Salmonella minnesota* R595 LPS to plasma or serum produces a marked decrease in the hydrated buoyant density of the LPS, as determined by CsCl isopycnic equilibrium density ultracentrifugation [23]. Either 10 μg of R595 or 20 μg of *E. coli* ^{125}I-LPS in 0.1 ml of NS was incubated with 0.3 ml of rabbit serum for 30 min. The density shifts obtained are shown in Figure 11, with approximately 50% of both 0111:B4 (d = 1.44 g/cm^3) and R595 (d = 1.38 g/cm^3) LPS being reduced to a density of less than 1.2 g/cm^3.

The role of lipids in producing this density shift was evaluated using delipidated plasma or serum. Delipidation was accomplished by extraction with butanol–diisopropyl ethanol (DIPE) [24]. The 0111:B4 or R595 ^{125}I-LPS, incubated at 37°C for 1 hr in either delipidated serum or plasma from human or rabbit sources (as shown in Fig. 12), demonstrates a descrease in density for R595 LPS from 1.38 to 1.34 g/cm^3, and for 0111:B4 LPS from 1.44 to a density partitioned at 1.42 and 1.34 g/cm^3. This is in marked contrast to results obtained using normal plasma or serum where both LPS demonstrate a shift to less than 1.2 g/cm^3.

Reconstitution of the delipidated serum was accomplished using various purified lipoproteins [25, 26] to determine if the ability to cause the LPS density shift could be restored with the purified serum lipoproteins. The ^{125}I-LPS was added to delipidated serum containing increasing concentrations of VLDL, LDL, or HDL from 20 to 200% of normal levels. Again the LPS–reaction mixture was incubated for 60 min at 37°C and ultracentrifuged on a CsCl gradient. The percentage of radioactivity banding at less than 1.2 g/cm^3 was determined for each condition. The results of adding increasing amounts of the various lipoproteins to the delipidated serum in producing density shifts for both R595 and 0111:B4 LPS are shown in Figure 13. As can be seen, the activity of delipidated serum can be reconstituted for both 0111:B4 and R595 LPS with HDL, but not LDL or VLDL. The concentration of HDL required is similar to normal levels and reconstitution of delipidated rabbit serum or plasma with human lipoproteins showed similar results. The percentage of ^{125}I-LPS at less than 1.2 g/cm^3 for either 0111:B4 or R595 LPS in delipidated serum plus HDL was similar to that seen in control experiments.

The ability of the various isolated lipoproteins to induce the density shift of ^{125}I-LPS to less than 1.2 g/cm^3 was also examined. After incubation for 60 min at 37°C in one of the lipoproteins at normal concentrations, the ^{125}I-LPS was ultracentrifuged in CsCl, and the percentage of LPS at d less than 1.2 g/cm^3 was measured. In the absence of delipidated serum, 0111:B4 LPS incubated with either VLDL or HDL produced a shift in density to less than 1.2 g/cm^3 for less than 12% of the LPS and, with LDL, 20% was found at d less than 1.2 g/cm^3. Using R595 LPS, 25% was shifted to d less than g/cm^3 by VLDL and HDL, whereas about 40% was shifted by LDL. Thus, it appears that nonlipid components found in the delipidated serum participate in the reduction in density for both LPS, although a partial direct interaction is also possible with the isolated lipoproteins.

To evaluate the ability of delipidated serum to alter the particle size of LPS, the sedimentation of parent R595 LPS and R595 LPS preincubated with delipidated serum was measured on 5–20% sucrose gradients, centrifuged at 45,000 rpm × 3 hr in an SW 50.1 rotor. Under these conditions greater than 75% of the parent ^{125}I-LPS was pelleted. In contrast, only 5% of the ^{125}I-LPS which

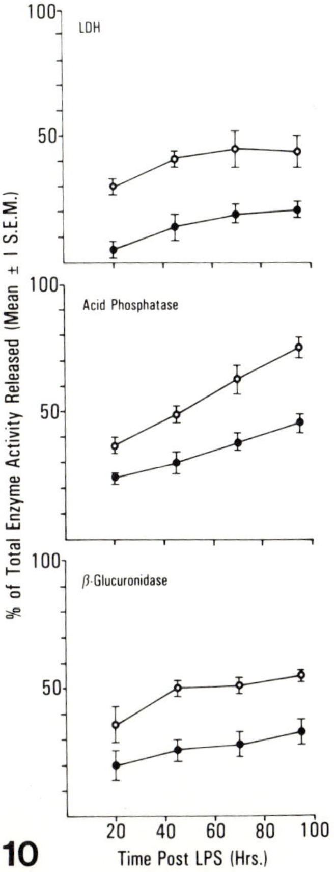

10

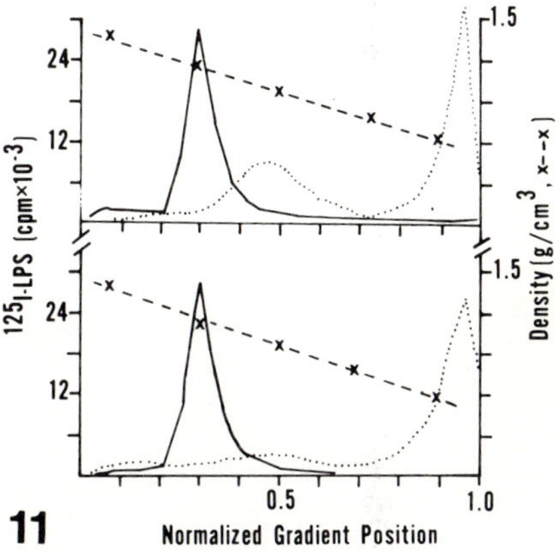

11

Fig. 10. Percentage of total LDH, acid phosphatase, and β-glucuronidase activity released by explanted H–Mφ in response to treatment with 10 μg/ml of R595 LPS. LPS-treated (-○-) and untreated (-●-) culture values represent the percentage of total activity (combined from supernatants and cell lysates) released into the media and are the mean ± 1 SEM for six experiments.

Fig. 11. The distribution of parent serum-altered 0111:B4 (a) and R595 (b) LPS in CsCl gradients after centrifugation at 40,000 RPM, 60 hr in an SW 50.1 rotor. Parent LPS is represented by a solid line, and LPS after exposure to normal human serum, by a dotted line. The density (g/cm^3) of gradient fractions is shown as a dashed line. The final reaction mixture consisted of 0.3 ml of serum and 0.1 ml of either 20 μg of 0111:B4 or 10 μg of R595 ^{125}I-LPS in normal saline.

was preincubated in delipidated serum was found in the pellet, and approximately 50% of the remaining ^{125}I-LPS banded in the upper fourth of the gradient in a homogeneous peak. Thus, the data suggest that a marked disaggregation of R595 LPS occurs in delipidated serum in the absence of all lipoproteins.

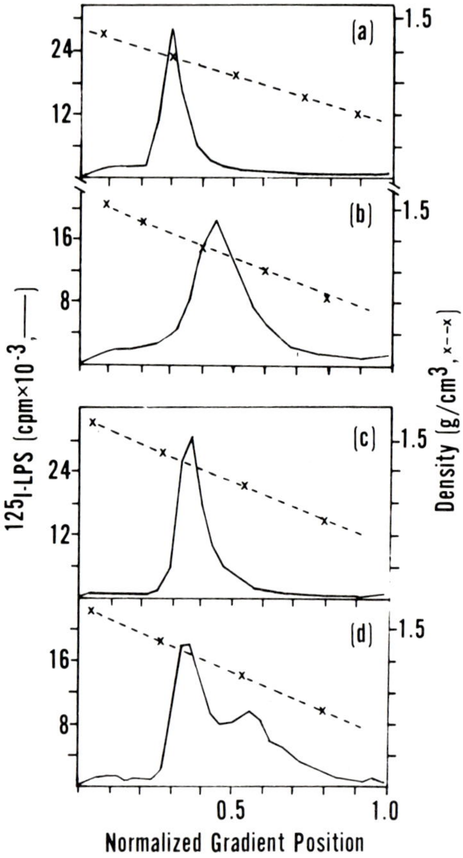

Fig. 12. The effect of delipidated serum on the distribution of LPS in CsCl gradients. (a) R595 LPS after exposure to 0.15 M NaCl, 20 mM EDTA; (b) R595 LPS after exposure to delipidated serum containing 20 mM EDTA; (c) 0111:B4 LPS after exposure to 0.15 M NaCl; and (d) 0111:B4 LPS after exposure to delipidated human serum are each shown by a solid line. The CsCl gradient was spun at 40,000 RPM for 60 hr in an SW 50.1 rotor, and in each panel the density (g/cm^3) of gradient fractions is shown as a dashed line. Refer to legend of Figure 11 for protocol.

Endotoxic Activity of Low-density LPS

Several assays, which reflect the immediate response of the host to LPS, were chosen to evaluate the endotoxic properties of LPS pretreated in normal serum or delipidated serum reconstituted with HDL. The aspects studied were the pyrogenic response, neutropenia induction, and anticomplementary activity.

Initially, the pyrogenic response produced by 50 ng of 0111:B4 LPS, preincubated with 1 ml of delipidated plasma or 1 ml of HDL (1.35 mg/ml), or 5 μg

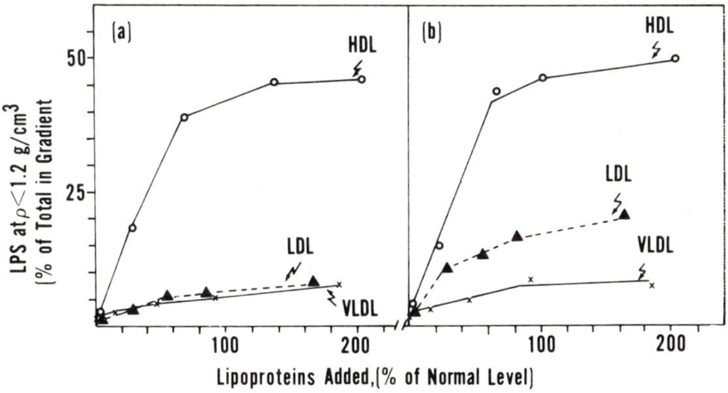

Fig. 13. Reconstitution of delipidated human serum with purified lipoproteins. The effect of delipidated serum–lipoprotein mixtures on the buoyant density of 0111:B4 LPS (a) and R595 LPS (b) is demonstrated.

of 0111:B4 LPS (d < 1.2 g/cm^3) isolated from a delipidated plasma–HDL mixture was studied. The minimal dose of native 0111:B4 LPS capable of producing a 1–1.5°C increase in core temperature was 50 ng [23]. The results demonstrate that neither delipidated plasma nor HDL alone are capable of inhibiting the pyrogenic activity. However, the LPS (d < 1.2 g/cm^3) from the delipidated plasma–HDL mixture was completely devoid of pyrogenic activity. Similar results are also obtained using R595 LPS.

LPS injection in rabbits induces a rapid, profound neutropenia [5, 23]. Using a similar protocol as in the pyrogenicity assay, the low-density LPS (d < 1.2 g/cm^3) was unable to produce the typical immediate neutropenia after the injection of 5 μg of material. In contrast, the injection of as little as 50 ng of 0111:B4 LPS in 1 ml of delipidated plasma reduced the circulating neutrophil count to less than 10% of pretreatment levels. Again, delipidated plasma was unable to inhibit this property of LPS.

An indirect but sensitive measure of the anticomplementary activity of R595 LPS is the rapid thrombocytopenia induced by injection into rabbits [27]. The decrease in platelets has been shown to be dependent upon intravascular complement activation with 50 μg of R595 LPS producing a 30–50% decrease in circulating platelet levels within 5 min. The ability of R595 LPS (d < 1.2 g/cm^3), altered by treatment with the delipidated plasma–HDL mixture, to induce thrombocytopenia was compared to native R595 LPS (d = 1.38 g/cm^3). A 250-μg dose of parent LPS was compared to 350 μg of low-density LPS, both given as an intravenous bolus. The altered LPS was unable to produce the immediate thrombocytopenia seen with parent LPS, which produced a 50% reduction in platelet counts.

Clearance and Tissue Distribution of Low-density LPS

To evaluate the potential conversion of LPS in vivo to the low-density form, plasma samples were obtained at various intervals after the iv injection of 250 μg of either 0111:B4 or R595 LPS in rabbits. Using isopycnic CsCl ultracentrifugation, greater than 90% of the LPS found in plasma was in the low-density form (d > 1.2 g/cm^3) by 30 min post injection. Thus, the decrease in clearance rates, as shown in Figure 1, was associated with the appearance of the low-density form of LPS in plasma.

To determine the clearance rates of low-density LPS, plasma was isolated from rabbits 1 hr after the injection of 250 μg of 0111:B4 LPS. Ten-milliliter aliquots of this plasma containing low-density LPS were injected into different rabbits, and blood clearance kinetics were followed. The concentration of low-density 0111:B4 LPS present at 1 min post injection was identical to that predicted from estimated blood volumes. Over 4 hr, approximately 50% of the initial dose was cleared by the liver and the remaining intravascular LPS, which was greater than 95% confined to the plasma phase, had a half-life of 15 hr. Similar data were obtained using dialyzed low-density R595 LPS isolated from CsCl gradients. Again, initial blood concentrations were identical to that predicted from blood volumes and the half-life of plasma retained LPS was about 15 hr.

The tissue distribution of low-density ^{125}I-R595 LPS was studied at 180 min post injection. As before with parent LPS, the greatest portion of tissue-bound LPS was found in the liver, which contained 65%. However, a marked difference was noted in the adrenal uptake, with an 18-fold higher concentration of low-density LPS, as compared to parent LPS levels. The remaining tissues studied, including liver, had LPS concentrations similar to those seen in parent LPS-treated animals at 180 min post injection. The use of low-density 0111:B4 LPS produced similar results with, again, three to four fold greater concentration in adrenal tissue above that seen in liver tissue.

DISCUSSION

These experiments examine several of the cellular and molecular interactions of LPS within the experimental host. Both the smooth LPS, *E. coli* 0111:B4 and the rough mutant LPS, *Salmonella minnesota* R595, were used in each experimental protocol. *E. coli* 0111:B4 LPS is highly water soluble, has a MW of about 1×10^6, consists of about 35% by weight of lipid A, and is an activator of the alternative complement pathway [4]. In contrast, R595 LPS is a large aggregate with a sedimentation coefficient of greater than 50S, is poorly water soluble, consists of greater than 90% by weight of lipid A, and activates the classical complement pathway [4]. However, even with these marked differences (particularly in levels in amount of biologically active lipid A) both LPS moieties appear to function similarly in each of the interactions examined.

Clearance and Tissue Distribution of LPS

The initial interaction of LPS with the blood and tissues of rabbits was studied using ^{125}I-radiolabeled LPS. The use of covalently labeled ^{125}I-LPS, which has been shown to have unaltered biophysical and biologic functions [6], has eliminated the problems seen with other labeling procedures using ^{32}P [28], ^{51}Cr [29], or ^{3}H [30] LPS.

After the iv injection of LPS, the clearance from the vascular compartment is noted to be biphasic. This is explained by both the rapid initial tissue localization of LPS as well as a modification of the LPS remaining in plasma with a prolonged half-life of greater than 10 hr. The initial rapid clearance from blood appears to be dependent upon the aggregate size of the LPS. The large macromolecular aggregates of R595 LPS (a 78S aggregate) is cleared much more efficiently than the smaller, 17S aggregate of 0111:B4 LPS. The markedly reduced clearance of LPS in the second phase is coincident with the formation of a low-density form of LPS, which will be discussed later.

For both types of LPS, the tissue distribution is similar after iv injection, with the liver demonstrating the localization of the majority (40–50%) of the tissue-bound LPS within 5 min post injection. This corresponds to the findings of others [28, 31, 32] demonstrating that the liver is the major site of uptake of LPS immediately after injection. However, the tissue concentrations of 0111:B4 LPS show a relative (but not absolute) increase in splenic concentration over liver and lung tissue at 5 min post injection, as compared to R595 LPS concentrations (Fig. 4, 5). The larger aggregates of R595 LPS allow more rapid uptake by the reticuloendothelial system (RES), both by circulating and tissue-fixed phagocytes. This increased uptake with subsequent induction of aggregation would account for the increased numbers of labeled neutrophils and monocytes noted histologically in alveolar capillaries in R595 LPS-treated animals. The disaggregation and clearance of neutrophil clumps would explain the subsequent decrease in concentration of R595 LPS found in lung tissue to levels similar for 0111:B4 LPS at 180 min post injection. Concomitantly, a decreased ability of the H–Mϕ to ingest 0111:B4 LPS due to its smaller aggregate size [33] would produce the relative higher concentration found in the more sensitive splenic Mϕ, as has been shown in the clearance of other particulate matter from blood (reviewed in [34]).

At both 5 and 180 min post injection, the H–Mϕ was shown to be the major site of localization for both 0111:B4 and R595 LPS with electron microscopic autoradiography demonstrating the ^{125}I-LPS concentrated within the phagocytic vacuoles of the H–Mϕ. The other major sites of localization, ie the spleen and lungs, also demonstrated the ^{125}I-LPS to be associated with the splenic Mϕ, and the monocytes and neutrophils trapped within splenic sinusoidal and alveolar capillary beds, confirming the major role of the RES in clearing the vascular compartment of LPS [31, 32, 34].

Production and Interaction of Low-density LPS

As stated earlier, the appearance in plasma of low-density LPS ($d < 1.2$ g/cm^3) coincides with the markedly reduced rate of clearance for LPS retained in plasma and appears to protect the LPS partially from phagocytic ingestion after the initial rapid clearance of parent LPS by the RES. Five minutes after injection, only 55% of the 0111:B4 LPS and 17% of the R595 LPS remained in the blood and greater than 95% of each was contained in the platelet-free plasma. The ability of the conversion of LPS to the low-density form to reduce the clearance rate of LPS is supported by the lack of an initial rapid clearance phase when either low-density 0111:B4 or R595 LPS was injected.

HDL has been shown to be the major participant in serum (or plasma) responsible for the reduction of the buoyant density of LPS to less than 1.2 g/cm^3. The addition of HDL in physiological concentrations to delipidated serum or plasma restored both the ability to cause the density shift in LPS to less than 1.2 g/cm^3 and also the capacity to abrogate the biologic activities of LPS responsible for the pyrogenic response, acute thrombocytopenia induction, and the anticomplementary action. In contrast, the additions of VLDL or LDL in concentrations up to twice physiologic values were ineffective in reconstituting normal serum (or plasma) effects.

The present studies also demonstrate that nonlipid components of serum or plasma facilitate the HDL–LPS interaction, possibly by disaggregating the LPS prior to interacting with HDL. However, the delipidated serum (or plasma) alone was unable to inhibit several LPS-dependent biologic responses. In the case of 0111:B4, the addition of HDL alone neither produced a density shift in a significant amount of the parent LPS nor inactivated the pyrogenic response in the absence of delipidated serum. Although each lipoprotein, HDL, LDL, or VLDL, was able to induce partially a density shift in parent R595 LPS, the addition of delipidated serum markedly reduced any effect by VLDL and LDL. Thus, in vivo, the interaction of LPS with lipoprotein appears to be specific for HDL. Again, despite the marked differences in physical characteristics of the two types of LPS employed, both types produce similar pathophysiologic alterations in vivo in the rabbit and both interact in a similar fashion with serum components to produce a density shift of the parent LPS to less than 1.2 g/cm^3. The data thus support the concept that LPS is the ubiquitous moiety of Gram-negative bacteria responsible for their pathophysiologic effects.

The induction of a density shift in LPS to d less than 1.2 g/cm^3 also produces an alteration in the tissue distribution with a selective uptake of the LPS by the adrenal glands. This is demonstrated by the increased concentration of 0111:B4 LPS (which undergoes an increased percentage shift to $d < 1.2$ g/cm^3 compared to R595 LPS) in the adrenals at 180 min post injection. Also the adrenal LPS concentrations were markedly increased up to fivefold over all other tissues for both LPS when the low-density form of each was injected. Autoradiographic evaluation at the light microscopic level has isolated the uptake to the cytoplasm of adrenal cortical cells. This is in contrast to all other tissues where localization appears confined to phagocytic cells. The mechanism responsible for the selective uptake of the LPS–HDL complex by the adrenal cortical cells may be the presence of high-affinity receptors for lipoproteins demonstrated on these cells [35]. This mechanism for selective uptake and the known direct cellular cytotoxicity of LPS [36] may contribute to the Waterhouse-Friderichsen syndrome seen in Gram-negative septicemia. Although adrenal injury per se probably does not contribute directly to endotoxemic shock, injury to adrenal function may compromise the host's ability to survive and recover from a septic insult [37].

Furthermore, although the density shift in LPS abrogates several of the acute host responses to LPS, the low-density form retains its toxicity in mouse lethality assays [23]. Therefore, the altered LPS must still be capable of interacting with the host's mediation systems, although, probably not complement. The major uptake of low-density LPS by the liver and spleen, albeit at a slower rate, indicates that cell-mediated host responses are still potentially involved. The injection of low-density LPS ($d < 1.2$ g/cm^3) into rabbits confirms that the ability to induce shock, and DIC is uninhibited. This is consistent with the results presented showing absence of major degradation of the glycolipid backbone of LPS or removal of bound radioiodine as determined by SDS–PAGE analysis of low-density LPS contained in platelet-free plasma 180 min post injection. Even LPS extracted from liver homogenates 3 hr after localization to the phagocytic vacuoles of the H–Mϕ shows no evidence of dissociation of radiolabel or disruption of the glycoprotein structure, again, as determined by SDS–PAGE analysis. Thus, although the H–Mϕ clears the majority of either circulating parent or low-density LPS forms, it appears that the H–Mϕ is incapable of degrading LPS [38]. The marked increase in amount of LPS localized in bile fluid at 3 hr post injection, with levels equivalent to that of liver tissue, suggests that the LPS is processed by the H–Mϕ and eliminated via the biliary tract. Whether the hepatocyte is also involved in this LPS processing is not known.

LPS Interaction with H–Mϕ

The H–Mϕ is the major site of localization of iv injected LPS and, as such, may play a central role in initiating the pathologic effects of LPS. The preparative technique developed permits the evaluation of the LPS interaction with a pure population of explanted H–Mϕ. The isolation technique produces recovery of H–Mϕ in numbers comparable to that achieved by others [39]. These cells have been well characterized as similar to other Mϕ populations using morphology, diffuse intense staining for nonspecific esterase activity, possession of Fc and C3b membrane receptors, and ability to phagocytose latex and zymosan particles avidly. Mϕ being end-stage cells, the absence of outgrowth of other cell types for up to 10 days in culture also confirms the purity of the culture population.

Using adherent peritoneal exudate cells (PEC), Gordon et al [40] have suggested that inducible enzymes in Mϕ are controlled by at least three different mechanisms which are dependent on type of stimulus. Some enzymes, such as plasminogen activator, are stimulated by both nonparticulate and particulate stimuli [21]. The acid hydrolases are also stimulated by particulate phagocytosis but also show a dose-dependent increased release with phagocytic load. While, in contrast, lysozyme is unaltered in both production and secretion by phagocytic challenge [22].

LPS has been used as a stimulant in several studies using PEC with a generalized increase in most enzymes studied except lysozyme, which is unaffected [22]. Using H–Mϕ, we have shown that treatment of explanted H–Mϕ with LPS produces a marked selective increase in the neutral proteinase, plasminogen activator; the cellular marker, LDH, and the constitutive enzyme, lysozyme. The changes in acid hydrolase levels studied were minimal. Although LDH, β-glucoronidase, and acid phosphatase all showed an increased release of enzyme activity, this appears to be a nonspecific effect of direct LPS-induced cellular toxicity and is paralleled closely by the loss of DNA from LPS-treated cultures. Thus, the response of H–Mϕ to LPS appears to be both a nonspecific toxicity response and a selective response with a marked increase and subsequent release of several cellular enzymes.

Several other investigations support the concept of LPS toxicity to Mϕ by showing transient depression of the phagocytic activity of the RES [41], direct histologic evidence of H–Mϕ damage [42], and toxicity for splenic and pulmonary Mϕ in in vitro organ culture [43] after treatment with LPS. This cytotoxicity appears to be a direct cellular effect of uptake of LPS and, in preliminary studies, we have been unable to block the effect using prostaglandin-synthesis inhibitors, superoxide scavengers, or steroidal suppression of neutral proteinase production. The LPS suppressive effect on the RES with direct cytotoxicity for H–Mϕ and the induction of production and release of mediators from H–Mϕ may combine to contribute to the effects of endotoxemia [42, 44]. In support of this are the histologic findings of inflammatory foci surrounding H–Mϕ and the concomitant release of multiple acid hydrolases into liver effluents during endotoxemia. The induction of tolerance to LPS in rabbits prevents these events. The central role of damage to the liver is also accentuated by the increased lethality of LPS injection after pretreatment with galactosamine, which selectively damages hepatocytes [45]. The early pathologic changes found in the liver with H–Mϕ swelling and disruption, clumping of platelets and neutrophils intravascularly, and marked deposition of fibrin can all be induced by H–Mϕ mediators, such as arachidonic acid metabolites and procoagulant substances. Greisman et al [46] report pyrogenic tolerance to LPS in rabbits by direct infusion of LPS into the portal vein. Tolerance to LPS may well represent a form of cellular tachyphylaxis occurring in the fixed Mϕ population of the host, thus preventing further release of systemically active mediators.

SUMMARY

Several of the cellular and molecular interactions of LPS within the experimental host have been examined in an attempt to elucidate the potential role of these interactions in initiating the pathophysiologic events of endotoxemia. Using ^{125}I-LPS, the clearance of LPS from the blood and its tissue, cellular and subcellular localization was established. The H–Mϕ was found to be the major host site of intravenous LPS localization and, subsequently, the effect of LPS on explanted H–Mϕ was demonstrated to be both directly cytotoxic and stimulatory of a selective increase in several cellular enzymes. Both the depression in H–Mϕ function and the stimulation of release of local and systemic mediators by LPS give the H–Mϕ a potential central role in initiating endotoxemic shock and DIC. Finally, the marked reduction in the clearance rate of LPS remaining in plasma after the initial rapid tissue localization was found to coincide with a density shift to less than 1.2 g/cm^3 for the parent LPS. This density shift was found to be dependent upon binding of the LPS to HDL in the serum or plasma and, with the presence of cellular HDL receptors, accounts for a shift in tissue localization to the adrenals. A postulated effect of direct adrenal damage by LPS can thus be invoked as contributing to the endotoxemic syndrome.

ACKNOWLEDGMENTS

The authors wish to thank Charles G. Cochrane for support and advice and Ms. Betty Goddard for her help in preparing the manuscript. This is Publication No. 2292 from the Department of Immunopathology, Scripps Clinic and Research Foundation, 10666 North Torrey Pines Road, La Jolla, CA 92037. This work was supported in part by Grants AI–07007, AI–15136, and GM–07037 from the National Institutes of Health. R.J. Ulevitch is the recipient of a Research Career Development Award No. AI–00391.

REFERENCES

1. Kiani D, Quinn El, Burch KH, Madhavan T, Saravolatz, Neblett TR: The increasing importance of polymicrobial bacteremia. JAMA 242:1044, 1979.
2. Centanni E: Untersuchungen uber das infektions-fieber-das Fiebergift der Bakterien. Dtsch Med Wochenschr 20:148, 1894.
3. Boivin A, Mesrobeanu L: Recherches sur les antigenes somatique et sur les endotoxines des bacteries. I. Considerations generales et expose des technique utilisees. Revue Immunol 1:553, 1935.
4. Morrison DC, Ulevitch RJ: The effects of bacterial endotoxins on host mediation systems. AM J Pathol 93:525, 1978.
5. Ulevitch RJ, Cochrane CG, Henson PM, Morrison DC, Doe WF: Mediation systems in bacterial lipopolysaccharide induced hypotension and disseminated intravascular coagulation. I. The role of complement. J Exp Med 142:1570, 1975.
6. Ulevitch RJ: The preparation and characterization of a radioiodinated bacterial lipopolysaccharide. Immunochemistry 15:157, 1978.
7. Mathison JC, Ulevitch RJ: The clearance, tissue distribution and cellular localization of intravenously injected lipopolysaccharide in rabbits. J Immunol 123:2133, 1979.

8. Maier RV, Ulevitch RJ: The response of isolated rabbit hepatic macrophages to lipopolysaccharide. Circ Shock, in press, 1981.
9. Ulevitch RJ, Johnston AR, Weinstein DB: New function for high density lipoproteins. Their participation in intravascular reactions of bacterial lipopolysaccharides. J Clin Invest 64:1516, 1979.
10. Galanos C, Luderitz O, Westphal O: A new method for the extraction of R lipopolysaccharides. Eur J Biochem 9:245, 1969.
11. Downer NW, Robinson NC, Capaldi RA: Characterization of a seventh different subunit of beef heart cytochrome *C* oxidase. Similarities between the beef heart enzyme and that from other species. Biochemistry 15:2930, 1976.
12. Rodgers AW: "Techniques of autoradiography." New York: Elsevier Publishing Co., 1969.
13. Card LG, VanTubergen RP: High resolution autoradiography. I. Methods. J Cell Biol 15:173, 1962.
14. Mills DM, Zucker-Franklin D: Electron microscopic study of isolated Kupffer cells. Am J Pathol 54:1147, 1969.
15. Lentz PE, DiLuzio NR: Isolation of adult rat liver macrophages (Kupffer cells). In Fletcher S (ed): "Methods in Enzymology," New York: Academic Press, 1974, p. 647.
16. Emeis JJ, Planque B: Heterogeneity of cells isolated from rat liver by Pronase digestion: Ultrastructure, cytochemistry, and cell culture. J Res 20:11, 1975.
17. Rotman B, Papermaster BW: Membrane properties of living mammalian cells as studied by enzymatic hydrolysis of fluorogenic esters. Proc Natl Acad Sci 55:134, 1966.
18. LePecq J-B, Paoletti C: A new fluorometric method for RNA and DNA determination. Anal Biochem 17:100, 1966.
19. Li CY, Lam KW, Yam LT: Esterases in human leukocytes. J Histochem Cytochem 21:1, 1973.
20. Crofton RN, Disselhoff-den Dulk, MCC, vanFurth R: The origin, kinetics and characteristics of the Kupffer cells in the normal steady state. J Exp Med 148:1, 1978.
21. Unkeless, JC, Gordon S, Reich E: Secretion of plasminogen activator by stimulated macrophages. J Exp Med 139:834, 1974.
22. Gordon S. Todd J. Cohn ZA: In vitro synthesis and secretion of lysozyme by mononuclear phagocytes. J. Exp Med 139:1228, 1974.
23. Ulevitch RJ, Johnston AJ: The modification of biophysical and endotoxic properties of bacterial lipopolysaccharides by serum. J Clin Invest 63:1313, 1978.
24. Cham BE, Knowles BR: A solvent system for delipidation of plasma or serum without protein precipitation. J Lipid Res 17:176, 1976.
25. Weinstein DB, Carew TG, Steinberg D: Uptake and degradation of low density lipoprotein by swine arterial smooth muscle cells with inhibition of cholesterol biosynthesis. Biochim Biophys Acta 424:404, 1976.
26. Miller NE, Weinstein DB, Carew TE, Koschinsky, Steinberg D: Interactions between high density and low density lipoproteins during uptake and degradation by cultured human fibroblasts. J Clin Invest 60:78, 1977.
27. Ulevitch RJ, Cochrane CG: Role of complement in lethal bacterial lipopolysaccharide-induced hypotensive and coagulative changes. Infect Immunol 19:204, 1978.
28. Howard JG, Rowley D, Wardlaw, AC: Investigations on the mechanism of stimulation of non-specific immunity by bacterial lipopolysaccharides. Immunity 1:181, 1958.
29. Braude AI, Carey FJ, Sutherland D, Zalesky M: Studies with radioactive endotoxin. I. The use of ^{51}Cr to label endotoxin of *Escherichia coli*. J Clin Invest 34:850, 1955.
30. Schrader WH, Woolfrey BF: Studies with tritiated endotoxin. I. Preparation and analysis of tritiated endotoxin. AM J Pathol 42:225, 1963.
31. Noyes HE, MCInturf CR, Blahuta GJ: Studies on distribution of *Escherichia coli* in mice. Proc Soc Exp Biol Med 100:65, 1959.

32. Chedid L, Parant F, Parant M, Boyer F: Localization and fate of ^{51}Cr-labeled somatic antigens of smooth and rough salmonellae. Ann NY Acad Sci 133:712, 1966.
33. Tate WJ. Douglas H, Braude AI: Protection against lethality of *E. coli* endotoxin with "O" antiserum. Ann NY Acad Sci 133:746, 1966.
34. Benacerraf B: Functions of the Kupffer cells. In Rouller CH (ed): "Liver Morphology, Biochemistry, Physiology," New York: Academic Press 1964, p 37.
35. Kovanen PT, Faust JR, Brown MS, Goldstein JC: Low density lipoprotein receptors in mouse adrenal cortex. I. Receptor-mediated uptake of low density lipoprotein and utilization of its cholesterol for steroid synthesis in cultured adrenocortical cells. Endocrinology 104:599, 1979.
36. Shands JW, Peavy DL, Gormus BJ, McGraw J: In vitro and in vivo effects of endotoxin on mouse peritoneal cells. Infect Immunol 9:106, 1974.
37. Thompson EB, Lippman ME: Mechanism of action of glucocorticoids. Metabolism 23:159, 1974.
38. Bona CA: Fate of endotoxin in macrophages: Biological and ultrastructural aspects. J Infect Dis 128:574, 1973.
39. Melly MA, Duke LJ, Koenig MG: Studies on isolated cultured rabbit Kupffer cells. J Res 12:1, 1972.
40. Gordon S, Unkeless JC, Cohn ZA: Induction of macrophage plasminogen activator by endotoxin stimulation and phagocytosis. Evidence for a two stage process. J Exp Med 140:995, 1974.
41. Biozzi C, Benacerraf B, Halpern BN: The effect of *Salmonella typhi* and its endotoxin on the phagocytic activity of the reticuloendothelial system in mice. Br J Exp Pathol 36:226, 1955.
42. Levy E, Path FC, Buebner BH: Hepatic changes produced by a single dose of endotoxin in the mouse. Light microscopy and histochemistry. Am J Pathol 51:269, 1967.
43. Heilman DH: The selective toxicity of endotoxin for phagocytic cells of the reticuloendothelial system. Int Arch Allergy 26:63, 1965.
44. Janoff A, Weissmann G, Zweifach BW, Thomas L: Pathogenesis of experimental shock. IV. Studies on lysosomes in normal and tolerant animals subjected to lethal trauma and endotoxemia. J Exp Med 16:451, 1962.
45. Galanos C, Freudenberg MA, Reutter W: Galactosamine induced sensitization to the lethal effects of endotoxin. Proc Natl Acad Sci 76:5939, 1979.
46. Greisman SE, Woodward CL: Mechanisms of endotoxin tolerance. VII. The role of the liver. J Immunol 105:1468, 1970.

Influence of Bacterial Endotoxins on Neutrophilic Leukocytes: Lack of a Correlation Between In Vivo and In Vitro Responses

Mark E. Wilson, Priscilla Munkenbeck, and David C. Morrison

Dramatic hemodynamic and hematologic aberrations are common manifestations of systemic bacterial infection (sepsis) due to Gram-negative organisms. Among the earliest hematologic changes associated with Gram-negative sepsis is a marked decline in the number of circulating leukocytes [1]. The disappearance of leukocytes from the circulation appears to result, in the main, from the margination and subsequent extravasation of neutrophilic leukocytes (also referred to as polymorphonuclear leukocytes, or PMNL) [2]. The "rebound" leukocytosis which often follows the initial leukopenic phase reflects a mobilization of neutrophil precursors from bone marrow reserves [3]. Although transient leukopenia and ensuing leukocytosis are frequently observed in patients with Gram-negative sepsis, and may in some instances be of diagnostic value (eg, in the diagnosis of sepsis neonatorum), the relationship (if any) between the egress of circulating neutrophils and the pathogenesis of septic shock remains ambiguous.

Bacterial endotoxins (lipopolysaccharide, or LPS) are an integral component of the outer cell wall of Gram-negative bacteria, and their capacity to interact with a number of cellular and humoral host mediator systems is well documented [4,5]. Intravenous administration of purified endotoxin into experimental animals induces pathophysiologic changes commonly observed during Gram-negative sepsis, including the aforementioned fluctuations in the level of circulating neutrophils [6,7]. The in vivo effects of LPS upon circulating neutrophils have fostered speculation that LPS–neutrophil interactions may be important in initiating certain pathophysiologic changes associated with septic shock. Neutrophils possess both oxygen-dependent and -independent mechanisms for effecting microbial killing [8] which may also mediate tissue injury during acute inflammation [9,10]. Activation of any or all of these mechanisms, resulting from the interaction of bacteria and/or bacterial products with circulating neutrophils,

Pathophysiological Effects of Endotoxins at the Cellular Level, pages 157–172

could have deleterious effects upon intravascular and extravascular tissues [11,12]. Indeed, intravenous injection of *Escherichia coli* endotoxin into rabbits produces a typical inflammatory response [7].

Present information regarding the capacity of bacterial endotoxins to interact with, and to subsequently alter the metabolic disposition of, neutrophils is limited and often contradictory. Endotoxins possess a high affinity for binding to mammalian cell membranes, including leukocytes [13,14]. The consequence of such binding upon neutrophil function has been examined using a number of in vitro assay techniques including bactericidal activity, oxygen metabolism, glucose metabolism, and lysosomal enzyme release. Some investigators have reported that endotoxins stimulate leukocyte respiration [15,16], whereas others are unable to confirm this finding [17,18]. Direct glucose oxidation via the hexose monophosphate shunt has also been reported to be enhanced in endotoxin-stimulated neutrophils [15,18–20]. Further, endotoxins have been reported to induce a modest release of lysosomal enzyme constituents from neutrophilic leukocytes [19,21]. In many of these studies, however, incubations were performed in the presence of serum, thus making it difficult to ascertain whether endotoxin stimulation of neutrophil metabolism was a direct effect or one mediated via endotoxin activation of the complement system. This is an important distinction in that it has been established that certain peptide fragments generated during complement activation, notably C5a, can stimulate neutrophil oxidative metabolism [22].

We have investigated the capacity of purified bacterial endotoxins isolated from both smooth (*E. coli* 0111:B4, *E. coli* 055:B5) and rough (*Salmonella minnesota* R595) strains to interact with, and to stimulate oxidative metabolism of, human and rabbit neutrophils isolated from peripheral blood. In preliminary studies, binding of endotoxin to human neutrophils was determined using ^{125}I-endotoxin prepared according to Ulevitch [23]. The same preparations of endotoxin employed in these studies have previously been shown to exert potent biologic activities, as assessed by lethality in mice [24], neutropenia-inducing activity in both rabbits [25] and rhesus monkeys [6] lymphocyte proliferation [26], macrophage activation [27], and reactivity in the *Limulus* amoebocyte lysate assay. Metabolic activity of control and endotoxin-treated neutrophils was determined by measuring superoxide dismutase-inhibitable cytochrome *c* reduction (reflecting superoxide anion formation), generation of a chemiluminescent response, and release of lysosomal enzymes into the extracellular medium.

Certain endotoxic preparations, in particular those extracted using the method of Boivin and Mesrobeanu [28], contain protein as well as lipopolysaccharide. Further extraction of this material with aqueous phenol yields a protein component which readily associates, via noncovalent interaction, with the lipid A portion of the lipopolysaccharide molecule [29]. This protein, termed lipid A-associated protein, or LAP, exerts profound biologic effects on lymphocytes [29,30], macrophages [31] and certain nonlymphoid cells [32]. We have, therefore, included in our investigation an examination of the effects of LAP upon human neutrophil function in order to ascertain whether certain disparities in the results obtained by various investigators using different endotoxin preparations may be due to variable contamination with this protein.

MATERIALS AND METHODS

Reagents

Dulbecco's phosphate-buffered saline (PBS), pH 7.4, was purchased from Grand Island Biological Company, NY, and was supplemented with 0.1% glucose. In some studies, Tris-buffered Tyrode's solution (TBSA), pH 7.2, containing 0.25% bovine serum albumin (Sigma Chemical Co., St. Louis, MO) was substituted. Phorbol myristate acetate (PMA), zymosan A, phenolphthalein glucuronide, and β-NADH were purchased from Sigma Chemical Company, St. Louis, MO. The divalent cation ionophore (A23187) was generously provided by Dr. R. L. Hamill of Eli Lilly and Company, Indianapolis, IN. Insoluble aggregates of human γ-globulin were prepared by Dr. E. Parks, Research Institute of Scripps Clinic, La Jolla, CA.

Rabbits

Male and female New Zealand white rabbits (2–2.5 kg) were obtained from Rancho de Conejo, Vista, CA. Animals were given water and food ad libitum prior to the beginning of each experiment.

Bacterial Endotoxins

LPS was extracted from *E. coli* 0111:B4 according to the method of Westphal and Jann [33] using the hot water–phenol procedure. The LPS was subsequently fractionated by Sepharose 4B (Pharmacia, Uppsala, Sweden) chromatography [34], yielding a polysaccharide-rich, high-molecular weight fraction designated as LPS I, and a lower molecular weight fraction, LPS II. The latter LPS has a lower polysaccharide content and, therefore, a higher percentage, by weight, of lipid A. LPS from another smooth strain, *E. coli* 055:B5, was extracted and chromatographed in an identical manner. Protein contamination in these LPS preparations was less than 1%. LPS from the rough mutant *S. minnesota* R595 was prepared by the phenol–chloroform–petroleum ether method described by Galanos et al [35]. Lipid A-associated protein (LAP) was isolated from butanol-extracted *E. coli* 0111:B4 LPS as previously described [36]. LPS preparations (and LAP) were suspended in normal saline and stored at −20°C. R595 LPS was routinely subjected to brief sonication prior to use.

Neutrophil Isolation

Peripheral venous blood was obtained from healthy adult volunteers (informed consent was given) by standard techniques. Blood was also obtained from the ear of healthy rabbits by means of a 19-gauge needle. In each case blood was collected into polypropylene tubes containing 1:6 (v/v) acid–citrate–dextrose. Neutrophils were isolated by 1g sedimentation in 2.0% (human) or 2.5% (rabbit) gelatin (Difco, Detroit, MI) in normal saline [37]. Residual erythrocytes were lysed in 0.83% ammonium chloride, pH 7.0 (pH adjusted with NaOH). The leukocyte pellets thus obtained were washed in PBS (human) or TBSA (rabbit) and then resuspended in buffer. PMNL were enumerated using a hemacytometer, and the cell suspensions were adjusted to 1–2 × 10^7 PMNL/ml of buffer.

Particle Opsonization

Twenty to forty milligrams of zymosan A were suspended in normal saline and placed in a boiling water bath for 30 min. The boiled zymosan was pelleted by centrifugation, resuspended in normal human or rabbit serum (either fresh or stored at −70°C) at a concentration of 4 mg/ml and incubated for 30 min at 37°C. The opsonized zymosan was centrifuged at 4°C, washed and resuspended in buffer at a stock concentration of 4 mg/ml. The stock zymosan suspension was stored at −20°C.

Chemiluminescence Assay

Chemiluminescent (CL) responses were monitored using a Beckman LS-230 liquid scintillation spectrometer, operated at ambient temperature in the out-of-coincidence mode as originally described by Allen et al [38]. Dark-adapted plastic liquid scintillation vials containing 10^7 PMNL in PBS were warmed to 37°C for 15 min under constant agitation in a Dubnoff metabolic shaker. The vials were sequentially transferred to the scintillation counter and monitored for 0.5-min intervals in order to determine background CL values. CL reactions were initiated via the addition of opsonized zymosan, LPS, LAP, or PBS alone. Where indicated, PMNL were preincubated with LPS or LAP prior to challenge with zymosan. CL responses were monitored for 0.5-min intervals at 1, 5, 10, 15 and 20 min after stimulus addition, and all vials were maintained at 37°C between counting intervals.

Superoxide Assay

Superoxide anion production by resting and stimulated human and rabbit PMNL was determined via the reduction of ferricytochrome *c*, as previously described [39]. Leukocytes (5 × 10^6 or 10^7 PMNL) were preincubated for 15 minutes at 37°C in 12 × 75-mm polypropylene tubes (Falcon Plastics, Oxnard, CA) in the presence or absence of superoxide dismutase, (50 μg/ml, Sigma Chemical Co., St. Louis, MO) and LPS, or LAP. Reactions were initiated via the addition of cytochrome *c* (0.08 mM, type VI from horse heart, Sigma Chemical Co., St. Louis, MO) simultaneously with LPS, LAP, zymosan, or PBS. Following a 20-min incubation period, the reactions were terminated by placing the tubes on ice. PMNL were pelleted by centrifugation at 400g for 5 min at 4°C. The supernatant fractions were diluted 1:1 with PBS and assayed spectrophotometrically for reduction of cytochrome *c* by measuring absorbance at 550 nm. Absorbance values were converted to nanomoles of cytochrome *c* reduced using the extinction coefficient for cytochrome *c* of $2.1 \times 10^4\ M^{-1}cm^{-1}$ [40], and corrected for that portion of cytochrome *c* reduction which was not specifically inhibitable by superoxide dismutase.

Enzyme Assays

The cytoplasmic enzyme lactic dehydrogenase (EC 1.1.1.27) was assayed by the method of Wroblewski and LaDue [41] and was employed as a marker for cell lysis. β-Glucuronidase (EC 3.2.1.31) activity was determined using phenolphthalein glucuronide as substrate, as previously described [42]. Total releasable enzyme activity was measured by disrupting the cells in 0.3% Triton X-100, and release was expressed as a percentage of total.

LPS-Induced Neutropenia

Catheters were inserted into the central ear artery of rabbits and firmly secured in place. After discarding the first few drops of blood, a 1-ml sample was withdrawn and retained for determination of preinjection PMNL counts. The catheter was flushed with heparinized saline (10 U/ml) in order to minimize clotting. One-milliliter suspensions of the various LPS preparations were administered intravenously (bolus injection), and 1-ml blood samples were withdrawn at specific intervals after LPS injection. PMNL were enumerated by standard techniques using a hemacytometer and values are expressed as a percentage of preinjection levels.

RESULTS

Effects of Endotoxins on PMNL In Vivo

LPS-induced neutropenia in rabbits. Bacterial endotoxins induce rapid and pronounced changes in the level of circulating neutrophils in a number of animal species (including man, rhesus monkeys, and rabbits). In order to confirm the in vivo biologic activity of the endotoxic preparations employed in the present studies, we examined the capacity of these endotoxins to induce neutropenia in rabbits. The results of this study are presented in Figure 1. Maximal or near-maximal depression of circulating neutrophil levels was observed within 10 min after iv administration of either *E. coli* 0111:B4 PII or *S. minnesota* R595 LPS. As little as 0.5 μg of either form of LPS was capable of inducing a significant neutropenic response, with R595 LPS exhibiting slightly greater potency. The capacity of the polysaccharide-deficient R595 LPS to induce such a response

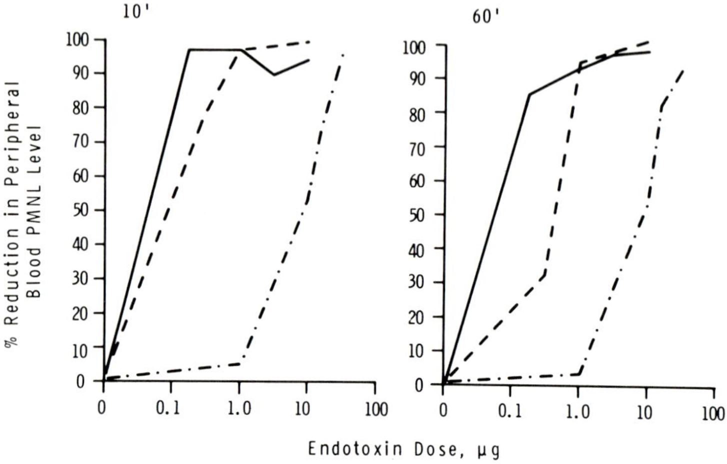

Fig. 1. The effect of intravenous injection of LPS or LAP on the levels of circulating polymorphonuclear leukocytes in rabbits. (Left panel) PMNL levels 10 min postinjection of *E. coli* 0111:B4 LPS (---), *S. minnesota* R595 LPS (———), or LAP (-·-). (Right panel) PMNL levels 60 min postinjection of LPS or LAP.

would suggest that the neutropenia-promoting activity of LPS may reside in the lipid A region.

The capacity of lipid A-associated protein to induce hematologic changes in rabbits was similarly examined. As seen in Figure 1, LAP was significantly less potent than either form of LPS in promoting a neutropenic response, with a dose of 10 µg inducing slightly more than 50% reduction in the circulating neutrophil level. The possibility cannot be excluded that this weak LAP response may in fact be due to minor LPS contamination. The data would suggest, nevertheless, that LAP has a minimal direct effect on circulating neutrophils.

Effects of Endotoxins on PMNL In Vitro

Influence of LPS on superoxide production by rabbit neutrophils. In view of the potent in vivo effects of LPS upon rabbit neutrophils, we then turned our attention toward an examination of the direct effects of these toxins upon PMNL function, employing a number of in vitro assay techniques. One of the earliest detectable changes in the functional status of neutrophils is a rapid increase in oxygen consumption. Consumed molecular O_2 is converted to hydrogen peroxide via a divalent reduction reaction. The intermediate formed during the univalent reduction of O_2 is superoxide anion (O_2^-), and is readily measured indirectly using the cytochrome *c* reduction method. We have examined the effect of LPS upon rabbit PMNL superoxide production in order to ascertain whether endotoxins directly stimulate PMNL oxygen metabolism.

TABLE I. Influence of Several LPS Preparations on Superoxide Production by Rabbit Neutrophils

Stimulus[a]	Superoxide production[b] (nanomoles of cytochrome *c* reduced/20′)[c]
None	2.6 ± 1.3 (4)
Serum-treated zymosan	60.6 ± 20.3 (4)
Phorbol myristate acetate	42.5 ± 17.1 (4)
Calcium ionophore (A23187)	11.6 ± 2.8 (3)
Aggregated HGG	15.0 ± 7.0 (3)
LPS [250 μg/ml]	
E. coli 055:B5	2.2 ± 0.5 (4)
E. coli 0111:B4	1.8 ± 0.5 (4)
S. minnesota R595	3.1 ± 1.6 (4)

[a][PMA] = 1 μg/ml; [A23187] = 1μM; [agg. HGG] = 250 μg/ml.
[b]Results expressed as the mean ± 1 SD. Number of experiments performed is indicated in parentheses.
[c]Each reaction tube contained 5 × 10^6 PMNL.

TABLE II. Effect of *S. minnesota* R595 LPS on Superoxide Production by Rabbit Neutrophils

Stimulus	Superoxide production[a] (nanomoles of cytochrome *c* reduced/20′)[b]
None	2.5 ± 1.0 (9)
Serum-treated zymosan (2 mg)	70.3 ± 20.3 (9)
R595 LPS	
0.25 μg/ml	2.1 ± 0.9 (5)
2.5 μg/ml	3.0 ± 0.9 (5)
25 μg/ml	3.6 ± 1.7 (5)
250 μg/ml	3.8 ± 1.6 (9)

[a]Results expressed as the mean ± 1 SD. Number of experiments performed is indicated in parentheses.
[b]Each reaction tube contained 5 × 10^6 PMNL.

In marked contrast to the effects of LPS on rabbit PMNL in vivo, in vitro incubation of PMNL with three different LPS preparations (at a concentration of 250 μg/ml) exerted no demonstrable influence on O_2^- production (Table I). The responses elicited by a variety of known particulate and soluble stimuli are provided for comparative purposes. The failure of LPS to stimulate O_2^- production was not a function of the concentration employed, in that LPS concentrations ranging from 0.25 to 250 μg/ml yielded identical results (Table II). The data strongly indicate that, although endotoxins exert dramatic effects upon rabbit

TABLE III. Effect of LPS on Superoxide Production by Human Neutrophils

Stimulus	Superoxide production[a] (nanomoles of cytochrome *c* reduced/20′)[b]
None	8.9 ± 0.2 (5)
Serum-treated zymosan (2 mg)	83.3 ± 18.7 (5)
S. minnesota R595 LPS	
0.025 μg/ml	10.1 ± 2.2 (5)
0.25 μg/ml	10.5 ± 1.6 (5)
2.5 μg/ml	10.3 ± 1.7 (4)
25 μg/ml	9.5 ± 1.4 (5)
E. coli 0111:B4 LPS	
0.025 μg/ml	8.8 ± 0.3 (2)
0.25 μg/ml	8.6 ± 0.3 (2)
2.5 μg/ml	10.6 ± 0.4 (2)
25 μg/ml	10.2 ± 0.1 (2)

[a]Results expressed as the mean ± SD. Number of experiments performed is indicated in parentheses.
[b]Each reaction tube contained 10^7 human PMNL.

PMNL in vivo, these toxins have little or no direct effect upon PMNL oxidative metabolic function.

Influence of LPS on superoxide production by human neutrophils. In order to determine whether the inability of rabbit PMNL to respond directly to LPS was unique to this animal species, we examined the capacity of LPS to elicit a metabolic response from human PMNL in vitro. We initially studied the effects of two LPS preparations (*E. coli* 0111:B4 and *S. minnesota* R595) on superoxide production by human PMNL (see Table III). Consistant with the results obtained using rabbit PMNL, neither form of LPS (over a 3-log range of concentrations) elicited a measurable increase in superoxide production by human PNML.

Effect of LPS on CL generation by unstimulated and phagocytosing human PMNL. Exposure of PMNL to a variety of particulate and soluble stimuli induces the generation of an electronic state(s) as manifested by a measurable CL response [38,43]. The observed CL response is reflective of, and dependent upon, the oxidative metabolic activity of these cells. The CL assay, therefore, constitutes a useful technique for monitoring the potential effects of various agents on PMNL oxidative metabolism. We have utilized the CL assay to study the effects of LPS upon both unstimulated ("resting") and phagocytosing human PMNL. The temporal CL responses of human PMNL exposed to *E. coli* 0111:B4 LPS or *S. minnesota* R595 LPS are shown in Figures 2 and 3, respectively. As may be noted in the right panel of each figure, neither LPS alone induced a measurable CL response.

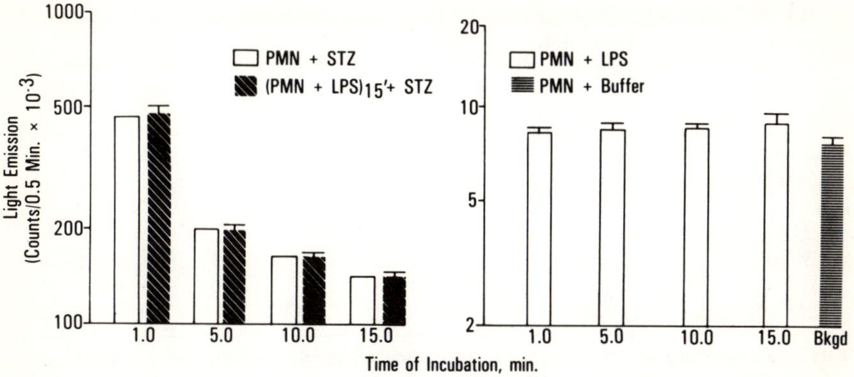

Fig. 2. The effect of *E. coli* 0111:B4 LPS on the temporal chemiluminescent responses of resting and phagocytosing human polymorphonuclear leukocytes. (Left panel) Effect of LPS pretreatment (0.25–25 μg/ml, 15 min at 37°C) on the phagocytosis-associated CL response elicited by serum-treated zymosan (STZ). (Right panel) Effect of LPS alone on the CL response of resting leukocytes.

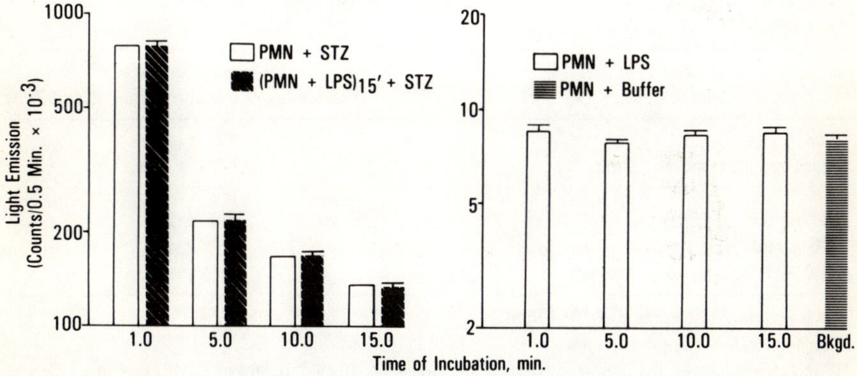

Fig. 3. The effect of *S. minnesota* R595 LPS on the temporal chemiluminescent responses of resting and phagocytosing human polymorphonuclear leukocytes. Experimental conditions were identical to those described in Figure 2.

Several investigators have reported variable effects of LPS upon neutrophil bactericidal activity, suggesting that LPS may modulate certain aspects of neutrophil function. In order to address this possibility, we have examined the effects of exposure of human PMNL to LPS upon the capacity of these cells to respond to a subsequent phagocytic challenge (left panel of Figs. 2 and 3). Incubation of PMNL with either LPS for 15 min prior to addition of a phagocytic stimulus (zymosan) was found to have no demonstrable effect upon the phagocytosis-

TABLE IV. Effect of *S. minnesota* R595 LPS on Lysosomal Degranulation by Resting and Phagocytosing Human Neutrophils

Stimulus	β-Glucuronidase release[a] (%)
None	2.0 ± 0.2[b]
Serum-treated zymosan (1 mg)	6.7 ± 0.6
LPS	
0.25 μg/ml	2.5 ± 0.7
2.5 μg/ml	2.4 ± 0.5
25 μg/ml	1.7 ± 0.3
(LPS + PMNL) $_{5'}$ → zymosan[c]	7.8 ± 1.5

[a]Results are expressed as a percentage of total releasable activity as determined by disruption of cells with Triton X-100. Following a 20-min incubation in the presence of stimuli, cells were removed by centrifugation. Aliquots of supernatants were assayed for β-glucuronidase by overnight incubation in the presence of phenolphthalein glucuronide substrate.
[b]Mean ± SD of three experiments, each performed in duplicate.
[c]PMNL incubated for 5 min at 37°C with LPS (33.3 μg/ml) prior to challenge with 1-mg of zymosan.

TABLE V. Effect of Lipid A-Associated Protein on Superoxide Production by Human Neutrophils

Stimulus	Superoxide production[a] (nanomoles of cytochrome *c* reduced/20′)
None	9.0 ± 24[b]
Serum-treated zymosan (1 mg)	85.9 ± 4.3
LAP	
1 μg/ml	8.9 ± 2.3
10 μg/ml	9.2 ± 1.6
(LAP + PMNL) $_{5'}$ → zymosan[c]	84.5 ± 5.3

[a]Each reaction tube contained 10^7 PMNL.
[b]Mean ± SD of three experiments.
[c]PMNL preincubated for 5 min at 37°C in the presence of 13.3 μg/ml of LAP prior to challenge with zymosan (1 mg).

associated CL response. Thus, LPS appears to exert neither a direct nor a modulatory effect upon neutrophil oxidative metabolism.

Effect of LPS on lysosomal degranulation by resting and phagocytosing human neutrophils. Neutrophils release certain lysosomal constituents into the surrounding medium following exposure to appropriate stimuli. This response is considered to be a secretory rather than a cytotoxic event in that the release of such constituents is not accompanied by the concomitant release of the cytoplasmic marker lactic dehydrogenase (LDH). Acid hydrolases present within lysosomes are capable of mediating tissue injury [9]. The ability

TABLE VI. Effect of Lipid A-Associated Protein on Lysosomal Degranulation by Resting and Phagocytosing Human Neutrophils

Stimulus	β-Glucuronidase Release[a] (%)
None	2.0 ± 0.3[b]
Serum-treated zymosan (1 mg)	5.0 ± 0.4
LAP	
1 μg/ml	2.5 ± 0.6
10 μg/ml	2.3 ± 0.3
(LAP + PMNL) $_{5'}$ → zymosan[c]	7.0 ± 1.9

[a]Refer to Table IV for experimental protocol.
[b]Mean ± SD of three experiments each performed in duplicate.
[c]PMNL preincubated for 5 min at 37°C with LAP (13.3 μg/ml) prior to challenge with zymosan (1 mg).

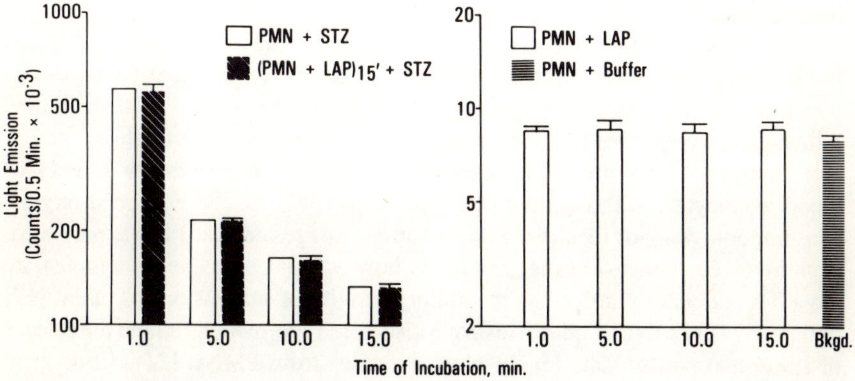

Fig. 4. The effect of lipid A-associated protein (LAP) on the temporal chemiluminescent responses of resting and phagocytosing human polymorphonuclear leukocytes. (Left panel) Effect of LAP pretreatment (1-10 μg/ml, 15 min at 37°C) on the phagocytosis-associated CL response elicited by serum-treated zymosan (STZ). (Right panel) Effect of LAP alone on the CL response of resting leukocytes.

of *S. minnesota* R595 LPS to induce release of the lysosomal marker β-glucuronidase from resting PMNL, or to modify enzyme release following addition of a phagocytic stimulus (zymosan), was studied. As may be seen in Table IV, R595 LPS (0.25–25 μg/ml) failed to induce β-glucuronidase release from unstimulated PMNL, nor did it significantly modify lysosomal discharge induced by zymosan. R595 LPS was not cytotoxic for PMNL, as determined by LDH release. Similar results were obtained using *E. coli* 0111:B4 LPS.

Effect of purified LAP on human neutrophil function. In view of the reported effects of lipid A-associated protein on a number of cell types, both lymphoid and nonlymphoid, we studied the interaction between LAP and human neutrophils in vitro. LAP (1–10 μg/ml) was determined to have no detectable effects upon neutrophil superoxide production (Table V), β-glucuronidase release (Table VI), or CL (Fig. 4). Further, preincubation of PMNL with LAP prior to challenge with a phagocytic stimulus did not alter the observed phagocytosis-associated metabolic response.

DISCUSSION

Bacterial endotoxins (lipopolysaccharides) exert profound in vivo effects upon a number of host mediator systems, both cellular and humoral. The polymorphonuclear leukocyte appears to be a primary target of circulating endotoxins. Endotoxins exhibit preferential binding to PMNL over other cellular components of blood (although appreciable binding to platelets does occur), and induce marked changes in the level of circulating PMNL in a number of animal species, including man.

Polymorphonuclear leukocytes, as inflammatory cells, can mediate cell and tissue injury via a host of oxidative and nonoxidative (principally enzymatic) mechanisms. The potential capacity of endotoxins to evoke PMNL-mediated inflammatory responses has been considered by numerous investigators. Strauss and Stetson [16] reported that endotoxins stimulate the respiratory activity of blood leukocytes. Although the cell type responsible for the enhanced oxygen consumption was not identified, these authors suggested that PMNL may have been involved. Other investigators have, however, been unable to demonstrate any effect of endotoxins on the respiratory activity of suspensions of rabbit [17] or human [18] PMNL. Endotoxins have also been reported to promote the release of lysosomal constituents (eg, β-glucuronidase) from PMNL [21]. Cline et al [44] emphasized the role of complement in certain endotoxin-induced metabolic changes in PMNL, suggesting that these bacterial products may not exert their effects upon PMNL directly.

In the present investigation, we have examined the capacity of purified lipopolysaccharides isolated from smooth and rough strains of Gram-negative bacteria to directly stimulate PMNL oxidative metabolism in the absence of serum. The endotoxic properties of these LPS preparations have been confirmed in this (Fig. 1) and previous studies. LPS concentrations as high as 250 μg/ml failed to stimulate superoxide production by rabbit or human PMNL. LPS similarly failed to augment the generation of chemiluminescence by human PMNL or to induce the release of the lysosomal enzyme β-glucuronidase. While this work was in progress, Proctor [18] reported similar results using a commercially available preparation of *E. coli* 0111:B4 LPS. Pretreatment of human PMNL with LPS prior to challenge with zymosan particles had no demonstrable effect upon phagocytosis-associated metabolic changes. Our results indicate that endotoxins, at concentrations in excess of those required to induce neutropenia in animals, exert neither a direct stimulatory nor a modulatory influence upon PMNL oxidative capacity.

Although endotoxins do not directly stimulate PMNL oxidative metabolism, the possibility remains that these bacterial products may, nevertheless, alter PMNL function indirectly. For example, endotoxin-induced activation of the complement system [45] can lead to the generation of peptide fragments which can directly stimulate PMNL metabolism [22]. Thus, the question of a participatory role of PMNL in the pathogenesis of endotoxic shock remains unanswered.

Endotoxins likely exist, both as integral cell wall constituents and upon release from Gram-negative bacteria during lysis, in noncovalent association with phospholipid and protein. The isolated protein component, termed lipid A-associated protein, exhibits biologic properties distinct from those of LPS [32]. We, therefore, examined the effects of this protein upon human PMNL function. Lipid A-associated protein, like LPS, was found to exert no demonstrable effect upon unstimulated or phagocytosing PMNL.

Phagocytosing PMNL exhibit an increased capacity to convert the water soluble dye nitroblue tetrazolium (NBT) to its reduced form, insoluble formazan. NBT reduction appears to be closely linked to PMNL oxidative metabolism and hexose monophosphate shunt (HMPS) activity [46]. Under normal aerobic conditions, NBT reduction by stimulated PMNL is considered to reflect the enhanced production of superoxide by these cells [47]. However, available evidence suggests that NBT reduction can occur via a nonoxidative mechanism(s) as well. Methylene blue stimulation of HMPS activity of PMNL from patients with chronic granulomatous disease (which lack the oxidase enzyme involved in the "respiratory burst" of phagocytosis) leads to significant NBT dye reduction [48].

Bacterial endotoxins are known to stimulate HMPS activity and NBT reduction in normal PMNL under conditions in which oxygen consumption and superoxide production are not altered [18]. It would appear, therefore, that endotoxin-induced NBT dye reduction by PMNL [18,49,50] is a nonoxidative event. The mechanism(s) by which endotoxins might promote NBT reduction and HMPS activation is not understood. Such a reaction might be initiated at the cell surface or via direct interaction with the enzymes and/or pyridine cofactors of the HMPS. Experiments are currently in progress to differentiate between these possibilities.

REFERENCES

1. Shubin H, Weil M, Nishijima H: Clinical features in shock associated with gram-negative bacteremia. In Urbaschek B, Urbaschek R, Neter E (eds): "Gram-Negative Bacterial Infections." New York: Springer-Verlag, 1975, p 411.
2. Athens JW, Haab OP, Raab SO, Mauer AM, Ashenbrucker H, Cartwright GE, Wintrobe MM: Leukokinetic studies. IV. The total blood, circulating and marginal granulocyte pools and the granulocyte turnover rate in normal subjects. J Clin Invest 40:989, 1961.
3. Mechanic RC, Frei E III, Landy M, Smith WW: Quantitative studies of human leukocytic and febrile response to single and repeated doses of purified bacterial endotoxin. J Clin Invest 41:162, 1962.

4. Morrison DC, Ulevitch RJ: The effects of bacterial endotoxins on host mediation systems. Am J Pathol 93:527, 1978.
5. Kabir S, Rosenstreich DL, Mergenhagen SE: Bacterial endotoxins and cell membranes. In Jeljaszewicz J, Wadstrom T (eds): "Bacterial Toxins and Cell Membranes." New York: Academic Press, 1978, p 59.
6. Ulevitch RJ, Cochrane CG, Bangs K, Herman CM, Fletcher JR, Rice CL: The effect of complement depletion on bacterial lipopolysaccharide (LPS)-induced hemodynamic and hematologic changes in the rhesus monkey. Am J Pathol 92:227, 1978.
7. Goodman ML, Way BA, Irwin JW: The inflammatory response to endotoxin. J Path 128:7, 1979.
8. Klebanoff SJ: Antimicrobial mechanisms in neutrophilic polymorphonuclear leukocytes. Semin Hematol 12:117, 1975.
9. Cochrane CG: Role of granulocytes in immune complex-induced tissue injuries. Inflammation 2:319, 1977.
10. Johnston RB Jr, Lehmeyer JE: Elaboration of toxic oxygen by-products by neutrophils in a model of immune complex disease. J Clin Invest 57:836, 1976.
11. Ryan GB, Majno G: Acute inflammation. Am J Pathol 86:185, 1977.
12. Handin RI, Karabin R, Boxer GJ: Enhancement of platelet function by superoxide anion. J Clin Invest 59:959, 1977.
13. Shands JW Jr: Affinity of endotoxins for membranes. J Infect Dis (Suppl) 128:S197, 1973.
14. Springer GF, Adye JC: Endotoxin-binding substances from human leukocytes and platelets. Infect Immun 12:978, 1975.
15. Graham RC Jr, Karnovsky MJ, Shafer AW, Glass EA, Karnovsky ML: Metabolic and morphologic observations on the effect of surface-active agents on leukocytes. J Cell Biol 32:629, 1967.
16. Strauss BS, Stetson CA Jr: Studies on the effect of certain macromolecular substances on the respiratory activity of the leucocytes of peripheral blood. J Exp Med 112:653, 1960.
17. Cohn ZA, Morse SI: Functional and metabolic properties of polymorphonuclear leukocytes. II. The influence of a lipopolysaccharide endotoxin. J Exp Med 111:689, 1960.
18. Proctor RA: Endotoxin in vitro interactions with human neutrophils: Depression of chemiluminescence, oxygen consumption, superoxide production, and killing. Infect Immun 25:912, 1979.
19. Bannatyne RM, Harnett NM, Lee K-Y, Biggar WD: Inhibition of the biologic effects of endotoxin on neutrophils by polymyxin B sulfate. J Infect Dis 136:469, 1977.
20. Buschmann H, Frohnwieser B: Factors influencing the phagocytosis-stimulated glucose oxidation in porcine leukocytes. Z Immun-Forsch 152:1, 1976.
21. Goldstein IM, Wünschmann B, Astrup T, Henderson ES: Effects of bacterial endotoxin on the fibrinolytic activity of normal human leukocytes. Blood 37:447, 1971.
22. Goldstein IM, Roos, D, Kaplan HB, Weissmann G: Complement and immunoglobulins stimulate superoxide production by human leukocytes independently of phagocytosis. J Clin Invest 56:1155, 1975.
23. Ulevitch RJ: The preparation and characterization of a radioiodinated bacterial lipopolysaccharide. Immunochemistry 15:157, 1978.
24. Curry BJ, Morrison DC: Role of complement in endotoxin initiated lethality in mice. Int J Immunopharmacology 1:125, 1979.
25. Ulevitch RJ, Cochrane CG, Henson PM, Morrison DC, Doe WF: Mediation systems in bacterial lipopolysaccharide-induced hypotension and disseminated intravascular coagulation. I. The role of complement. J Exp Med 142:1570, 1975.

26. Chiller JM, Skidmore BJ, Morrison DC, Weigle WO: Relationship of the structure of bacterial lipopolysaccharides to its function in mitogenesis and adjuvanticity. Proc Natl Acad Sci (USA) 70:2129, 1973.
27. Niemetz J, Morrison DC: Lipid A as the biologically active moiety in bacterial endotoxin-initiated generation of procoagulant activity by peripheral blood leukocytes. Blood 49:947,1977.
28. Boivin A, Mesrobeanu L: Recherches sur les antigènes somatiques et sur les endotoxines des bactéries. I. Considérations générales et exposé des techniques utilisées. Rev Immunol (Paris) 1:553, 1935.
29. Betz SJ, Morrison DC: Chemical and biologic properties of a protein-rich fraction of bacterial lipopolysaccharides. I. The in vitro murine lymphocyte response. J Immunol 119:1475, 1977.
30. Morrison DC, Betz SJ, Jacobs DM: Isolation and characterization of the fraction of LPS responsible for mitogenesis of C3H/HeJ spleen cells. J Exp Med 144:840, 1976.
31. Doe WF, Yang D, Morrison DC, Betz SJ, Henson PM: Macrophage stimulation by bacterial lipopolysaccharides. II. Evidence for independent differentiation signals delivered by lipid A and by a protein-rich fraction of LPS. J Exp Med 148:557, 1978.
32. Morrison DC, Wilson ME, Raziuddin S, Betz SJ, Curry BJ, Oades Z, Munkenbeck P: Influence of lipid A-associated protein on endotoxin stimulation of nonlymphoid cells. In Schlessinger D (ed): "Microbiology, 1980." Washington, D.C.: American Society for Microbiology, 1980, p 30.
33. Westphal O, Jann K: Bacterial lipopolysaccharide extraction with phenol-water and further applications of the procedure. In Whistler RL (ed): "Methods in Carbohydrate Chemistry." New York: Academic Press, 1965, p 83.
34. Morrison DC, Leive L: Fractions of lipopolysaccharide from *Escherichia coli* 0111:B4 prepared by two extraction procedures. J Biol Chem 250:2911, 1975.
35. Galanos C, Lüderitz O, Westphal O: A new method for the extraction of R lipopolysaccharides. Eur J Biochem 9:245, 1969.
36. Morrison DC, Betz SJ, Jacobs DM: Isolation of a lipid A bound polypeptide responsible for "LPS-initiated" mitogenesis of C3H/HeJ spleen cells. J Exp Med 144:840, 1976.
37. Henson PM: The immunologic release of constituents from neutrophil leukocytes. I. The role of antibody and complement on nonphagocytosable surfaces or phagocytosable particles. J Immunol 107:1535, 1971.
38. Allen RC, Stjernholm RL, Steele RH: Evidence for the generation of an electronically excited state(s) in human polymorphonuclear leukocytes and its participation in bactericidal activity. Biochem Biophys Res Commun 47:679, 1972.
39. Babior BM, Kipnes RS, Curnutte JT: Biological defense mechanisms. The production by leukocytes of superoxide, a potent bactericidal agent. J Clin Invest 52:741, 1973.
40. Massey V: The microestimation of succinate and the extinction coefficient of cytochrome *c*. Biochim Biophys Acta 34:255, 1959.
41. Wroblewski F, LaDue JS: Lactic dehydrogenase activity in blood. Proc Soc Exp Biol Med 90:210, 1955.
42. Talalay R, Fishman WH, Huggins C: Chromogenic substrates. II. Phenolphthalein glucuronic acid as a substrate for the assay of glucuronidase activity. J Biol Chem 166:757, 1946.
43. Trush MA, Wilson ME, VanDyke K: The generation of chemiluminescence by phagocytic cells. In DeLuca M (ed): "Bioluminescence and Chemiluminescence." Methods in Enzymology, vol 57, New York: Academic Press, 1978, p 462.
44. Cline MJ, Melmon KL, Davis WC, Williams HE: Mechanism of endotoxin interaction with human leucocytes. Brit J Haematol 15:539, 1968.
45. Kane MA, May JE, Frank MM: Interactions of the classical and alternate complement pathways with endotoxin lipopolysaccharide. J Clin Invest 52:370, 1973.

46. Karnovsky ML: The metabolism of leukocytes. Semin Hematol 5:156, 1968.
47. Baehner RL, Boxer LA, Davis J: The biochemical basis of nitroblue tetrazolium reduction in normal human and chronic granulomatous disease polymorphonuclear leukocytes. Blood 48:309, 1976.
48. Humbert JR, Gross GP, Vatter AE, Hathaway WE: Nitroblue tetrazolium reduction by neutrophils: Biochemical and ultrastructural effects of methylene blue. J Lab Clin Med 82:20, 1973.
49. Matula G, Paterson PY: Spontaneous in vitro reduction of nitroblue tetrazolium by neutrophils of adult patients with bacterial infection. N Eng J Med 285:311, 1971.
50. Park BH, Good RA: NBT test stimulated. Lancet 2:616, 1970.

Possible Contribution of Platelets to the Pathophysiology of Host Responses to Endotoxin

Richard I. Walker and J. Raymond Fletcher

Platelets are important cells in the inflammatory response of the host to microorganisms. For example, platelets interact with pathogens and release mediators of inflammation, participate in trapping microorganisms for phagocytes, and, although usually not bactericidal themselves, enhance the bactericidal effects of leukocytes [1–4]. This potentially beneficial response of platelets to infection can become detrimental if thrombosis becomes too extensive to be resolved effectively. The host response to microorganisms can also be induced to self-destructive levels by challenge with purified endotoxin, a lipid–polysaccharide component of the cell walls of Gram-negative bacteria.

Endotoxin can interact with phospholipids found in biological membranes [5] such as those of leukocytes and platelets, which contain glycerophosphatides with strong affinities for endotoxin [6]. The significance of platelet affinities for endotoxin is demonstrated by the finding that within minutes of injection, almost all labeled endotoxin is associated with the platelets [7, 8]. Later most of the endotoxin is found in the liver, an event which is consistent with the demonstrated particle transport role of platelets [9–12]. Platelet–particle complexes must pass the capillary network of nonreticuloendothelial (RE) organs to reach RE tissues such as the liver [10]. This process depends on the maintenance of a balance between aggregation and disaggregation of platelets.

Excessive damage to platelets and other defensive elements responsible for clearance of microbial agents by RE organs described above can impair the effective removal of the toxin, or the inflammatory process it induces, from the microcirculation. Several lines of evidence suggest that such damage to microcirculatory targets contributes to the pathophysiology of subsequent host responses to endotoxin. First, when equal doses of endotoxin are injected either directly into the hepatic circulation (thereby being cleared without first going through the microcirculation) or into the systemic circulation, mortality and

Pathophysiological Effects of Endotoxins at the Cellular Level, pages 173–185

hepatic injury are obtained only via the latter route [13, 14]. Second, when low-level, continuous endotoxin is administered, death is associated with persistance of small amounts of endotoxin in the circulation but without accumulation of amounts of the toxin in the liver which would normally be associated with a lethal challenge [9].

The apparent importance of microcirculatory events in the host response to endotoxin indicated the need for studies to evaluate further some aspects of the platelet–endotoxin interaction. The specific questions to be addressed by these studies were: (1) is microcirculatory damage associated with the lethal effects of endotoxin, (2) is the platelet–endotoxin interaction a dose-dependent phenomenon, (3) are primate platelets affected by endotoxin, and (4) which biochemical requirements are necessary to effect platelet responses to endotoxin.

MATERIALS AND METHODS

Experimental Animals

New Zealand white rabbits weighing approximately 2.5 kg were utilized in most experiments with platelet aggregation and in all experiments involving tolerant animals. In vitro experiments with dog platelets were conducted with cells obtained from mongrel dogs, but in vivo observations of platelet and leukocyte changes were made in beagles. Studies of genetic effects on microcirculatory clearance following challenge with endotoxin utilized an endotoxin-sensitive (C3HeB/FeJ) and an endotoxin-resistant (C3H/HeJ) strain of mice obtained from Jackson Laboratories [16].

Preparation of Platelet-Rich and Platelet-Poor Plasma

Heparinized blood was obtained by cardiac puncture of rabbits under metofane anesthesia. Platelet-rich plasma (PRP) containing about 850,000 platelets/mm^3 was obtained by centrifugation for 10 min at 300g [15]. Platelet-poor plasma (PPP) was obtained by centrifuging blood at 1200g for 10 min. Venous blood from baboons was treated similarly.

One-milliliter samples of PRP were placed in 10-mm polystyrene cuvettes containing a Teflon-coated magnetic stirrer. Aggregating agents were added to the cuvettes with a Hamilton syringe. Increased transmission (aggregation) was recorded from a Beckman DU-2 spectrophotometer equipped with a magnetic stirrer and a 37°C water-heated jacket [15].

Endotoxin Preparations

Salmonella typhi (0901) and *Escherichia coli* (0128:B12) lipopolysaccharides (Westphal preparations) were purchased from Difco Laboratories. Irradiated *E. coli* endotoxin was generously supplied by Dr. Lorand Bertok (Budapest, Hungary). Triethylamine (TEA) and sodium salts of lipid A from a Difco preparation of *E. coli* (0128.B12) endotoxin were kindly furnished by Dr. Lutz A. Kiesow (Bethesda, MD).

Dog platelet-aggregation studies were performed with endotoxin from *Serratia marcescens* (supplied by Dr. A. Nowotny, Philadelphia, PA). This preparation caused reproducible, dose-responsive aggregation of platelets. Excessive variations were obtained with other preparations.

Induction of Tolerance to Endotoxin

Rabbits were made tolerant to *S. typhi* endotoxin by receiving intraperitoneal injections of the toxin on 5 consecutive days (0.2, 0.3, 0.4, 0.5, and 0.5 mg, respectively). The efficacy of this treatment was determined in rabbits by the intravenous injection of a normally lethal dose of the toxin (0.15 mg/g).

Preparation of Antisera and Serological Tests

Rabbit antiserum which was primarily reactive against O-antigenic determinants of *S. typhi* was prepared in the manner described by Sippel et al [18]. Lipid A antiserum was obtained by subcutaneously injecting rabbits with a mixture of 50 mg of lipid A and 100 mg of N-acetyl-muramyl-L-alanyl-D-isoglutamine (Calbiochem) suspended in Freund's incomplete adjuvant (Difco). Immune serum was obtained 30 days following antigen injection.

Antibodies to *S. typhi* O-antigens were measured with an enzyme-linked immunosorbent assay (ELISA) to be described elsewhere [19]. Precipitin activities against lipid A were detected by the double-immunodiffusion test performed on agarose (Miles Laboratories).

Histological Studies of Mouse Lungs

The pulmonary microvasculature of mice challenged intraperitoneally (ip) with 1 mg of *S. typhi* endotoxin was studied in animals which had been sacrificed in a carbon dioxide chamber and whose lungs were inflated with a dual-purpose fixative before removal from the thoracic cavity [17]. These specimens were sectioned and stained for examination by light microscopy. Microcirculatory congestion was evaluated by counting the number of polymorphonuclear leukocytes in three 400 × fields in each of six sections taken from various parts of the lungs of each mouse. Four mice were examined at each time period studied after challenge with endotoxin.

RESULTS AND DISCUSSION

Persistance of Inflammatory Complexes in Microcirculation

If microcirculatory events are significant in determining survival during endotoxemia, animals which are resistant to endotoxin should have adaptations which favor resolution of inflammation at this site. To test this hypothesis, mice with normal sensitivity to endotoxin (C3Heb/FeJ) and others with genetic resistance to the toxin (C3H/HeJ) were challenged ip with 1 mg of *S. typhi* endotoxin [17]. This dose killed most sensitive mice within 24 hr, but none of the genetically resistant animals died.

When numbers of circulating leukocytes and platelets were followed after challenge with endotoxin, similar degrees of thrombocytopenia and leukopenia were observed over a 24-hr period. However, when the presence of inflammatory complexes containing leukocytes and, presumably [20], platelets in the pulmonary microcirculation was observed, distinct differences between the responses of C3Heb/FeJ and C3H/HeJ mice were seen (Fig. 1). At 1 hr after challenge, pulmonary capillaries in both groups of mice contained similar numbers of polymorphonuclear leukocytes, but by 3 hr after challenge these cells were decreasing in number in the capillaries of the resistant mice. By 6 hr after challenge the capillaries in C3H/HeJ mice were normal in appearance. In contrast, inflammatory complexes persisted at high levels in pulmonary capillaries of the sensitive mice for at least 12 hr. At this time, the C3Heb/FeJ mice were moribund.

Microcirculatory inflammation has been described by Urbaschek [20, 21] for endotoxin-challenged animals. He observed an early platelet aggregation and accumulation of inflammatory complexes in the microcirculation. This process was accompanied by damage to endothelial cells [21]. In surviving animals microvascular inflammation was resolved, but, in animals receiving a lethal challenge, it was not.

Platelet aggregates can damage microcirculatory capillaries by obstruction and liberation of vasoconstrictor substances [22]. Platelet aggregates also release cationic proteins capable of inducing vascular lesions [23]. Endothelial damage resulting from endotoxin-induced platelet release and other inflammatory factors in the microcirculation would further enhance aggregatory processes through local exposure of platelets to collagen. The significance of these aggregates is indicated by the report [24] that ADP-induced aggregation can cause a shock syndrome similar to that induced by endotoxin. As with endotoxin challenges, rabbits injected with ADP died with platelet aggregates in the microcirculation. In addition, it has been shown that unless platelet aggregation is blocked by treatment with inhibitory drugs, rabbits can be killed by the injection of preformed antigen–antibody complexes [8].

The experiments described above indicate that microcirculatory damage might be associated with the lethal effects of endotoxin. Furthermore, evidence is available in the literature to show that some of the microcirculatory damage seen after challenge with endotoxin could be due to platelet injury.

Induction of Platelet Damage by Endotoxin

Thrombocytopenia during endotoxemia can be due to direct and indirect effects of the toxin. Direct interaction of endotoxin with subprimate platelets can

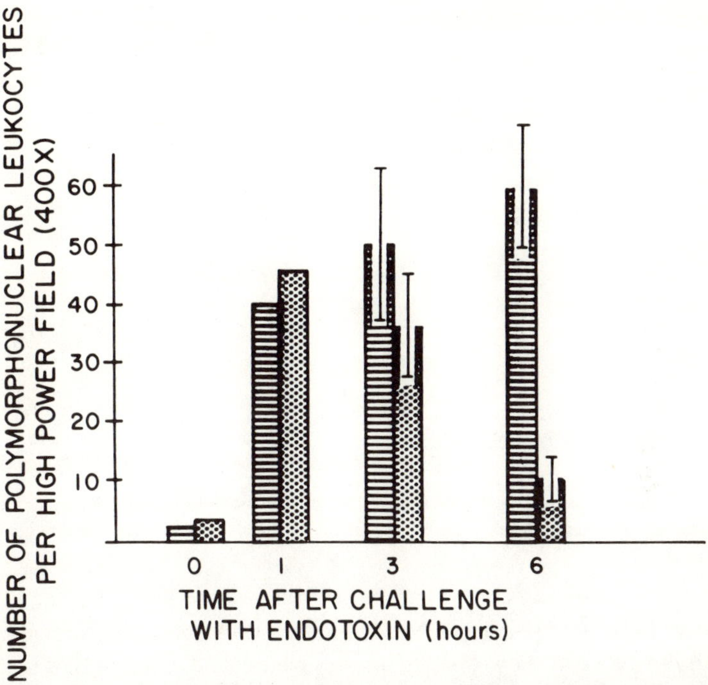

Fig. 1. Comparison of the numbers of polymorphonuclear leukocytes present (mean ± SD) in the pulmonary capillaries of genetically resistant (C3H/HeJ) (bar with dots) and normally sensitive (C3Heb/FeJ) (striped bar) mice at various times after challenge intraperitoneally with 1 mg of *S. typhi* endotoxin.

induce an aggregation response. When endotoxin was added to dog PRP in amounts increasing from 25 ng to 5 μg/ml [17], a dose-dependent increase in platelet aggregation was seen (Fig. 2). It is of interest that conventional antiaggregating agents including prostacyclin, the most potent antiaggregating agent yet discovered, failed to inhibit endotoxin-induced platelet aggregation in the dog [17]. The in vitro dose-responsiveness of dog platelets to endotoxin is consistent with previous in vivo observations [25, 26] that dogs challenged with small (4 μg/kg) doses of endotoxin had no observable thrombocytopenia, but those challenged with large sublethal (1.8 mg/kg) or lethal (2.75 mg/kg) doses experienced severe thrombocytopenia (the numbers of these cells in the circulation dropped from 400,000 to 40,000 cells/mm^3 in 15 min). In contrast, similar degrees of leukopenia were obtained in these dogs, regardless of whether the animals were challenged with small or large doses of endotoxin.

The dose-dependent nature of the direct effects of endotoxin on platelets was recently illustrated in another manner. Washida [27] noted that, at least in rabbits,

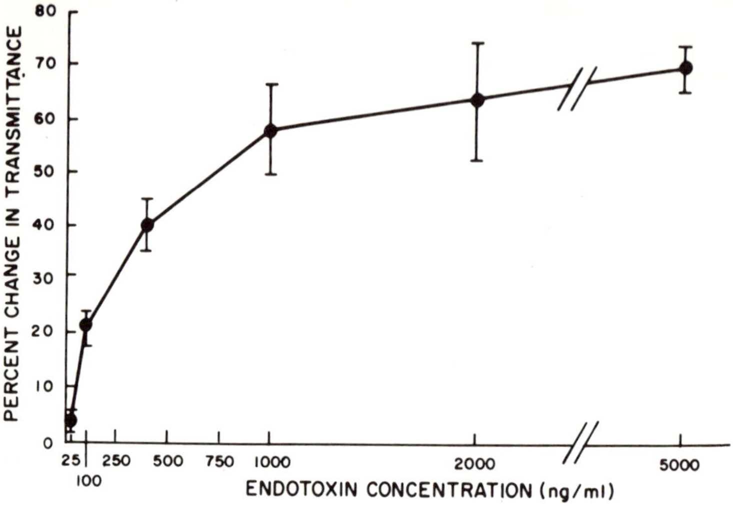

Fig. 2. Induction of platelet aggregation in canine platelet-rich plasma with increasing amounts of *S. marcescens* endotoxin. Each value represents the mean of 17 individual experiments with 17 different dogs.

the binding indexes of platelets for endotoxin showed a positive correlation with sensitivity to the toxin. Low-binding indexes were found in platelets of immature rabbits, and animals at this age were more resistant to endotoxin than were older rabbits, whose platelets had greater affinities for endotoxin.

Direct effects of endotoxin on primate platelets have been difficult to demonstrate in vitro, although human platelets do have receptors for endotoxin [28]. For example, aggregation responses readily observed in rabbit or dog PRP exposed to endotoxin are not obtained in primate preparations. However, thrombocytopenia does occur in primate models challenged with endotoxin (unpublished data). Male baboons weighing 20–30 kg were restrained and subjected to an LD_{50} dose of *E. coli* endotoxin (3 mg/kg given as an intravenous bolus via a peripheral vein). By 15 min after challenge, circulating platelets in those animals dropped to approximately 50% of normal values (Table I). Furthermore, when the responsiveness of the remaining platelets to ADP (2 μg/ml) was tested, they were found to be refractory to induction of aggregation.

Taken together, these experiments indicate that damage to platelets by endotoxin can be a direct dose-related phenomenon in subprimate animal models. Primate platelets are also affected in endotoxin-challenged animals, but in vivo studies do not distinguish whether these effects of endotoxin are direct or indirect.

Humoral Factors Affecting the Endotoxin–Platelet Interaction

Complement. The endotoxin–platelet interaction takes place in a plasma environment, in which various inorganic and organic humoral factors can affect

TABLE I. Thrombocytopenia and Platelet Responsiveness in Baboons Challenged with *E. coli* Endotoxin*

		Time after challenge (min)				
Test[a]	Baseline	15	60	120	180	240
Platelet count (× $10^3/\mu l$)	250 ± 20	169 ± 38	167 ± 66	150 ± 53	143 ± 69	117 ± 57
Responsiveness to ADP-induced aggregation[b]	100	10 ± 2	12 ± 2	8 ± 2	13 ± 4	18 ± 3

*A single intravenous LD_{50} dose (3 mg/kg) was administered to an unanesthetized animal.
[a]Results are expressed as mean ± SD for 10 animals.
[b]Data represent the percentage of change in transmission of light after addition of ADP (2 μg/ml) to baboon platelet-rich plasma.

its outcome. The importance of complement in mediating the platelet–endotoxin interaction was demonstrated by Brown and Lachmann [29]. In contrast to normal rabbits, animals depleted of C3-9 did not experience thrombocytopenia following injection of endotoxin, and all depleted animals survived. Furthermore, in C6-deficient animals, platelets were destroyed to a variable degree, and death could be associated with the extent of platelet injury.

Endotoxin causes aggregation of rabbit platelets in vitro. However, when the toxin is separated into its lipid and polysaccharide subcomponents, aggregation is no longer achieved [19]. Addition of 5 μg of either TEA or sodium salt preparations of lipid A to rabbit PRP did not induce aggregation, although aggregation of the same platelet preparation was accomplished with a comparable amount of the whole toxin [15]. Morrison [30] demonstrated that endotoxin preparations such as lipid A activate the classical complement pathway and do not cause platelet lysis. Activation of the alternative complement pathway by the polysaccharide portion of endotoxin is also required to obtain platelet lysis.

Evidence that both components of endotoxin are necessary for lysis or aggregation of platelets is shown by endotoxin-detoxification experiments. Polymyxin B binds to lipid A, the toxic moiety of the endotoxin complex, thereby detoxifying the endotoxin preparation [32]. This detoxified preparation also loses its platelet lytic capability. Similarily megarad doses of ^{60}Co destroy fatty acid groups on lipid A (L. Bertok, personal communication) and, thereby, reduce the toxicity of the molecule [32]. We found (unpublished data) that irradiated endotoxin no longer induces platelet aggregation.

Nonspecific factors. Factors other than complement may also influence the platelet–endotoxin interaction. For example, in studies of morphologic changes in human platelet membranes exposed to endotoxin, Ausprunk and Das [33] observed that plasma inhibits the endotoxin interaction with platelets. The nature of these plasma factors is not known. However, high-density lipoproteins, which can inhibit endotoxin interaction with erythrocyte membranes, have been detected in plasma [34]. In addition, humoral detoxifying factors are also present in plasma [35–37], and these substances could modify endotoxin effects on cellular targets. The detoxification capacity of plasma is greater than normal in animals made tolerant to endotoxin [38] and in those that are genetically unresponsive to its toxic effects [39].

Specific antibodies. Platelet-aggregation responses were studied in rabbits made tolerant to endotoxin [15, 19, 26]. By 7 days after the first injection to induce tolerance, the aggregation patterns of rabbit platelets exposed to endotoxin were significantly changed (Fig. 3). Time, from exposure to toxin to initiation of aggregation, was significantly shortened in PRP from tolerant animals and, in contrast to normal rabbits, aggregation was at least partly reversible. The induction of "tolerant" responses to endotoxin did not affect the aggregation responses of rabbit platelets exposed to ADP or collagen [40]. These characteristics were due to a humoral factor rather than a different platelet population, because cell-free plasma from tolerant animals could transfer tolerant responsiveness to normal rabbit platelets. The time at which this humoral factor appeared in the plasma (1 week after the first injection used to induce tolerance) correlated with the appearance of antipolysaccharide (but not antilipid A) antibodies in the plasma. Furthermore, when specific anti-O antisera was added to normal PRP, tolerant-like responses were observed following addition of endotoxin (Fig. 3). Addition of antilipid A did not accelerate aggregation or disaggregation after exposure of platelets to complete endotoxin, but it did appear to decrease the aggregation process.

Recently Bult and Herman [41] obtained responses with guinea pig platelets which were similar to those described above for rabbits made tolerant to the toxin. They found that pretreatment of guinea pigs with endotoxin caused a more rapid onset of the aggregation response when endotoxin was subsequently added to PRP from these animals. As with the rabbits, this effect was due to the presence of a humoral factor directed against the polysaccharide component of the endotoxin.

The role of specific antibody to polysaccharide components of endotoxin is unclear. Other factors may be responsible for survival of tolerant rabbits following an otherwise lethal challenge with endotoxin. Zaldivar and Scher [42] presented evidence that the presence of antibody titers, at least in mice, does not contribute to survival from challenge with endotoxin. It is possible that the tolerant response of platelets is important in the elimination of Gram-negative bacteria, but is of less consequence in protecting animals against the lethal effects of purified endotoxin.

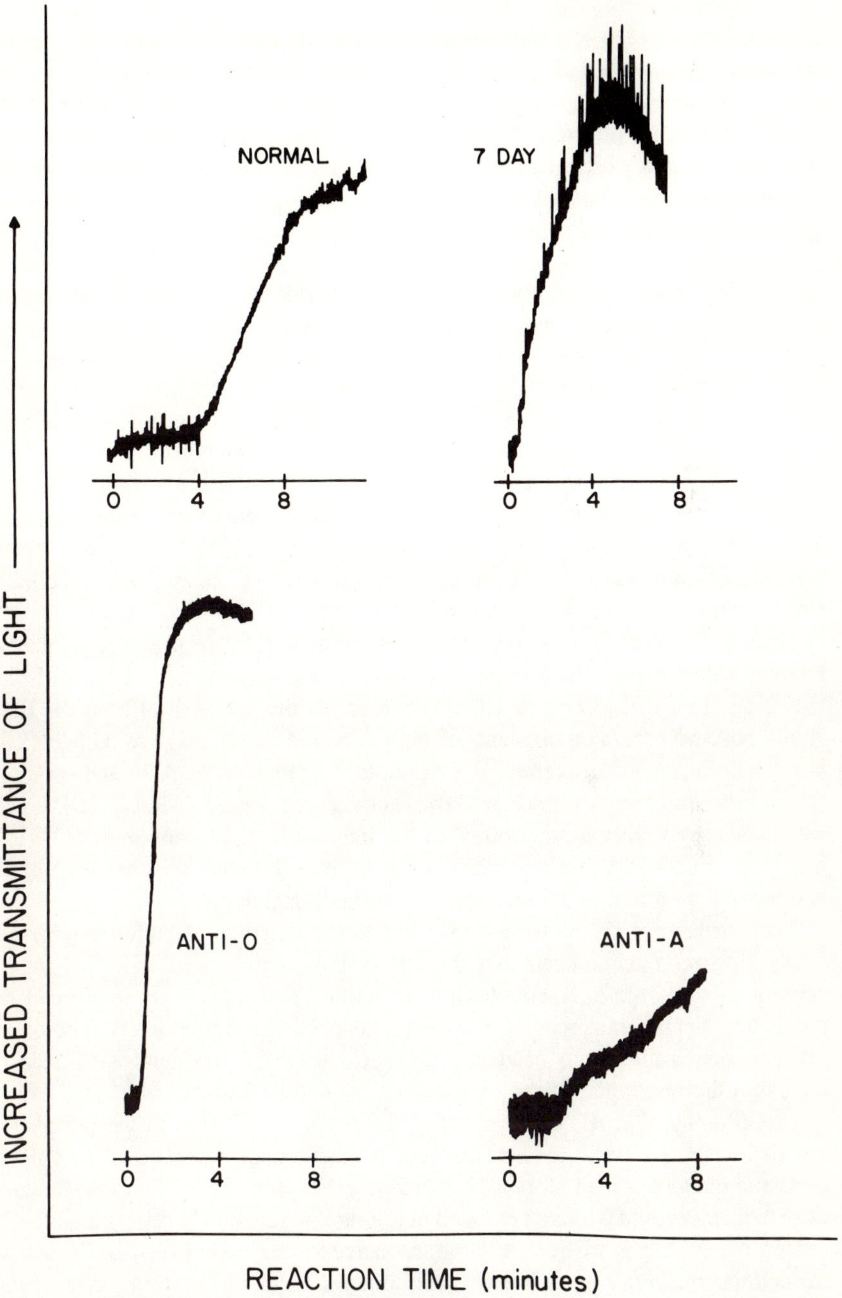

Fig. 3. Platelet aggregation in platelet-rich plasma taken from normal and 7-day tolerant rabbits (top). Aggregation patterns in lower half of the figure show reactions of normal platelet-rich plasma to endotoxin after addition of 0.2 cm^3 of specific antisera to lipid A or O-antigens. All preparations were challenged with 10 μg of *S. typhi* endotoxin at Time 0.

In summary, both complement and the lipid A portion of endotoxin seem to be required for the lethal effects of the toxin to be obtained. However, visible platelet damage (eg, aggregation of lysis) also requires the presence of the polysaccharide portion of the endotoxin molecule. Several humoral factors are present in plasma which may reduce damage associated with platelet–endotoxin interactions.

PERSPECTIVE

The host–endotoxin interaction must be considered as a sequential process involving both humoral and cellular elements rather than as isolated reactions of specific components of the inflammatory response. Persisting inflammatory activity in the microcirculation can be associated with sensitivity to the lethal effects of endotoxin. Platelets play an important role in the microcirculatory events of endotoxemia and dose-dependent endotoxin-induced damage to these cells may be a pivotal event in determining the extent of the response of other inflammatory components (eg, polymorphonuclear leukocytes). Endotoxin can be modified by humoral factors in plasma, and this altered endotoxin, in the presence of other humoral cofactors, can interact with platelets, which facilitate toxin removal from the bloodstream by RE cells.

Direct endotoxin damage to platelets may be altered by the presence of plasma factors, a possibility which requires further study. Aggregation or lysis of platelets, due to direct effects of endotoxin requires the presence of both the lipid A and polysaccharide components of the toxin. Indirect damage to platelets can also occur during endotoxemia. For example, factors such as ADP and collagen which are present in damaged capillaries activate platelets. Recently, ADP levels were associated with susceptibility to the lethal effects of endotoxin [43]. Although thrombocytopenia occurs in primates, it is not clear whether this injury is due more to direct or indirect damage to the platelets.

Platelet damage seems to be important to the outcome of endotoxemia, but it has long been recognized that the RE system is intimately involved in the consequences of the host–endotoxin interaction. This concept is reinforced by the finding that the transplant of lymphoreticular cells from genetically endotoxin-resistant donor mice into irradiated normally sensitive recipients significantly increased the resistance of the recipient animals to endotoxin [44].

It is possible that RE cells influence the reactivity of cells in the peripheral circulation to endotoxin. Mouse peritoneal macrophages are known to secrete platelet-aggregating mediators [45], which in excess may be detrimental. Release of such factors could be much reduced in tolerant or genetically resistant animals, as their macrophages are generally unresponsive to endotoxin [39, 46]. However, the cellular reactivity to endotoxin and humoral factors in the microcirculation also appears to be important in determining survival to endotoxin. This concept is supported by reports that the injection of otherwise toxic doses of endotoxin directly into the portal vein safely removes endotoxin from the circulation and prevents the lethal effects obtained via challenges with the same doses administered systemically [13, 14].

The importance of lymphoreticular cells in resistance to endotoxin may be explained if it could be shown that these cells secrete humoral factors which reduce platelet damage during endotoxemia. Although this idea is still speculative, it is strengthened by the work of Spillert [47] who demonstrated that a splenic factor which reduces thrombogenic activity enhances survival following challenge with endotoxin. Thus, although the mechanism(s) responsible for endotoxin lethality are still incompletely understood, considerable evidence is accumulating that microcirculatory events involving platelets contribute significantly to the pathophysiology of the host responses to endotoxin.

ACKNOWLEDGMENTS

This work was supported by the Naval Medical Research Development Command, NNMC, Department of the Navy, Research Task Numbers MPN06.01,0041 and ZF51524013.1030.

The opinions and assertions contained herein are the private ones of the writers and are not to be construed as official or reflecting the views of the Navy Department or the naval service at large.

The experiments conducted herein were conducted according to the principles set forth in the "Guide for the Care and Use of Laboratory Animals," Institute of Laboratory Resources, National Research Council, DHEW, Pub. No. (NIH) 74-23.

REFERENCES

1. Smith SB: Platelets in host resistance: In vitro interaction of platelets, bacteria, and polymorphonuclear leukocytes. Blut 25:104, 1972.
2. Wilder MS, Lubin DJ: Platelet enhancement of postphagocytic destruction of *Listeria monocytogenes* in cultured mouse peritoneal macrophages. Infect Immun 8:843, 1973.
3. Clawson CC: Modification of neutrophil function by platelets. In Baldini MG, Ebbe S (eds): "Platelets: Production, Function, Transfusion, and Storage." New York: Grune and Stratton, Inc, p 287, 1974.
4. Clawson CC, Rao GHR, Shite JG: Platelet interaction with bacteria. IV. Stimulation of the release reaction. Am J Pathol 81:411, 1975.
5. Shands JW, Jr: The physical structure of bacterial lipopolysaccharides. In Weinbaum G, Kadis S, Ajl SJ (eds): "Microbial Toxins, IV. Bacterial Endotoxins." New York: Academic Press, p 127, 1971.
6. Adye JC, Springer GF: Binding of endotoxin (LPS) by glycerophosphatides from human platelets and leukocytes. Naturwissenschaften 64:150, 1977.
7. Braude AI. Absorption, distribution and elimination of endotoxins and their deviatives. In Landy M, Braun W (eds): "Bacterial Endotoxins." New Jersey: Rutgers University Press, p 98, 1964.
8. Evans G. Fate of endotoxin in the circulation. Adv Exp Med Biol 23:81, 1971.
9. Noyes HE, McInturf CR, Blahuta GJ. Studies on the distribution of *Escherichia coli* endotoxin in mice. Proc Soc Exp Biol Med 100:65, 1959.

10. Van Aken WG, Vreeken J. Accumulation of macromolecular particles in the reticuloendothelial system (RES) mediated by platelet aggregation and disaggregation. Thromb Diath Haemorrh 22:496, 1969.
11. Donald KJ. The role of platelets in the clearance of colloidal carbon from blood in rabbits: A light and electron microscope study. Pathology 4:295, 1972.
12. Das J, Schwartz AA, Folkman J. Clearance of endotoxin by platelets: Role in increasing the accuracy of the limulus gelation test and in combating experimental endotoxemia. Surgery 74:235, 1973.
13. Rutenburg S, Skarnes R, Palmerio C, Fine J. Detoxification of endotoxin by perfusion of liver and spleen. Proc Soc Exp Biol Med 125:455, 1967.
14. Mori K, Matsumoto K, Gaus H. On the *in vivo* clearance and detoxification of endotoxin by lung and liver. Ann Surg 177:159, 1973.
15. Walker RI, Shields LJ, Fletcher JR. Platelet aggregation in rabbits made tolerant to endotoxin. Infect Immun 19:919, 1978.
16. Walker RI, Parker GA: Clearance of inflammatory cells from the microcirculation of endotoxin-resistant C3H/HeJ mice. Canad J Microbiol 26:725, 1980.
17. Fletcher JR, Ramwell PW: The effects of prostacyclin (PGI_2) on endotoxin shock and endotoxin-induced platelet aggregation in dogs. Circ Shock 7:299, 1980.
18. Sippel JE, Mamay HK, Weiss E, Joseph SW, Beasley WJ: Outer membrane protein antigens in an enzyme-linked immunosorbent assay for *Salmonella* enteric fever and meningococcal meningitis. J Clin Microbiol 7:372, 1978.
19. Walker RI, Beasley WJ: Evidence that anti-polysaccharide antibodies alter platelet responses to endotoxin in tolerant rabbits. Canad J Microbiol 26:1241–1246, 1980.
20. Urbaschek B: Addendum—The effects of endotoxins in the microcirculation. In Kadis S, Weinbaum G, Ajl SJ (eds): "Microbial Toxins V Bacterial Endotoxins." New York: Academic Press, p 261, 1971.
21. Urbaschek B, Urbaschek R, Neuhof H: Hemodynamic and metabolic changes in endotoxic shock in the miniature pig. Circ Shock 6:197, 1979.
22. Blaisdell FW, Lim RC Jr, Stallone RJ: The mechanism of pulmonary damage following traumatic shock. Surg Gynec Obstet 130:15, 1970.
23. Jorgenson L, Hovig T. Korsell HC, Mustard JF: Adenosine diphosphate-induced platelet aggregation and vascular injury in swine and rabbits. Amer J Pathol 61:161, 1970.
24. Evans G, Lewis AF, Mustard JF: The role of platelet aggregation in the development of endotoxin shock. Brit J Surg 56:624, 1969.
25. Walker RI, French JE, Walden DA, MacVittie TJ, Parker GA, Sobocinski PZ, Appelbaum FR: Protection of dogs from lethal consequences of endotoxemia with plasma or leukocyte transfusions. Advan Shock Res 4:89, 1980.
26. Walker RI: Platelet–endotoxin interactions in normal and resistant animals. In Eaker D, Wadstrom T (eds): "Natural Toxins." New York: Pergamon Press, in press.
27. Washida S: Endotoxin receptor site. II. Specificity of endotoxin receptor of platelets and sensitivity to endotoxin in vivo. Acta Med Okayama 256:125, 1978.
28. Hawiger J, Hawiger A, Timmons S: Endotoxin-sensitive membrane component of human platelets. Nature (London) 256:125, 1975.
29. Brown DL, Lachmann PJ: The behaviour of complement and platelets in lethal endotoxin shock in rabbits. Int Arch Allergy 45:193, 1973.
30. Morrison DC, Oades ZG: Mechanisms of lipopolysaccharide-initiated rabbit platelet responses. II. Evidence that lipid A is responsible for binding of lipopolysaccharide to the platelet. J Immunol 122:753, 1979.
31. Morrison DC, Jacobs DM: Binding of polymyxin B to the lipid A portion of bacterial lipopolysaccharides. Immunochemistry 13:813, 1976.

32. Previte JJ, Chang Y, El-Bisi HM: Detoxification of *Salmonella typhimurium* lipopolysaccharide by ionizing radiation. J Bacteriol 93:1607, 1967.
33. Ausprunk OH, Das J: Endotoxin-induced changes in human platelet membranes: Morphologic evidence. Blood 51:487, 1978.
34. Freudenberg MA, Bog Hansen TC, Back U, Jirillo E, Galanos C: Interaction of lipopolysaccharides with plasma high density lipoprotein in rats. Toxicon 17(S):52, 1979.
35. Skarnes RC: Host defense against bacterial endotoxemia: Mechanism in normal animals. J Exp Med 132:300, 1970.
36. Ulevitch RJ, Johnston AR: The modification of biophysical and endotoxic properties of bacterial lipopolysaccharides by serum. J Clin Invest 62:1313, 1978.
37. Johnson KJ, Ward PA, Goralnick S, Osborn MJ: Isolation from human serum of an inactivator of bacterial lipopolysaccharide. Am J Pathol 88:559, 1977.
38. Moreau SC, Skarnes RC: Host resistance to bacterial endotoxemia. Mechanisms in endotoxin-tolerant animals. J Infect Dis 128(S):122, 1973.
39. Sultzer BM, Goodman GW: Characteristics of endotoxin-resistant low-responder mice. In Schlessinger D (ed): "Microbiology 1977." Washington, DC: American Society for Microbiology, p 304, 1977.
40. Walker RI, Fletcher JR, Walden DA: Participation of granulocytes and humoral factors in resolution of platelet aggregates during endotoxemia. Experientia 36:255, 1980.
41. Bult H, Herman AG: Thromboxane A_2 biosynthesis during endotoxin-induced aggregation of platelets from normal and sensitized guinea pigs. Agents Act 9:560, 1979.
42. Zaldivar NM, Scher I: Endotoxin lethality and tolerance in mice: Analysis with the B-lymphocyte-defective CBA/N strain. Infect Immun 24:127, 1979.
43. Jabs CM, Ferrell WJ, Robb HJ: Plasma changes in endotoxin and anaphylactic shock (ATP, ADP and creatine phosphorus). Ann Clin Lab Sci 9:121, 1979.
44. Michalek SM, Moore RN, McGhee JR, Rosenstreich DL, Mergenhagen SE: The primary role of lymphoreticular cells in the mediation of host responses to bacterial endotoxin. J Infect Dis 141:55, 1980.
45. Blumenthal KM, Rourke FJ, Wilder MS: Platelet activation by cultured mouse peritoneal macrophages. J. Reticuloendo Soc 24:247, 1980.
46. Snyder SL, Eklund SK, Walker RI: Release of β-glucuronidase from peritoneal macrophages of normal and endotoxin-tolerant mice. Canad J Microbiol 25:1245, 1979.
47. Spillert CR, Lazaro EJ, Parmer LP, Jain KM: Protective effect of a splenic extract in mice with endotoxemia. J Trauma, in press.

Index

PROGRESS IN CLINICAL AND BIOLOGICAL RESEARCH

Vol 24: **Twin Research,** Proceedings of the Second International Congress on Twin Studies, Walter E. Nance, *Editor*
Published in 3 Volumes:
Part A: Psychology and Methodology
Part B: Biology and Epidemiology
Part C: Clinical Studies

Vol 25: **Recent Advances in Clinical Oncology,** Tapan A. Hazra and Michael C. Beachley, *Editors*

Vol 26: **Origin and Natural History of Cell Lines,** Claudio Barigozzi, *Editor*

Vol 27: **Membrane Mechanisms of Drugs of Abuse,** Charles W. Sharp and Leo G. Abood, *Editors*

Vol 28: **The Blood Platelet in Transfusion Therapy,** Tibor J. Greenwalt and G.A. Jamieson, *Editors*

Vol 29: **Biomedical Applications of the Horseshoe Crab (Limulidae),** Elias Cohen, *Editor-in-Chief*

Vol 30: **Normal and Abnormal Red Cell Membranes,** Samuel E. Lux, Vincent T. Marchesi, and C. Fred Fox, *Editors*

Vol 31: **Transmembrane Signaling,** Mark Bitensky, R. John Collier, Donald F. Steiner, and C. Fred Fox, *Editors*

Vol 32: **Genetic Analysis of Common Diseases: Applications to Predictive Factors in Coronary Disease,** Charles F. Sing and Mark Skolnick, *Editors*

Vol 33: **Prostate Cancer and Hormone Receptors,** Gerald P. Murphy and Avery A. Sandberg, *Editors*

Vol 34: **The Management of Genetic Disorders,** Constantine J. Papadatos and Christos S. Bartsocas, *Editors*

Vol 35: **Antibiotics and Hospitals,** Carlo Grassi and Giuseppe Ostino, *Editors*

Vol 36: **Drug and Chemical Risks to the Fetus and Newborn,** Richard H. Schwarz and Sumner J. Yaffe, *Editors*

Vol 37: **Models for Prostate Cancer,** Gerald P. Murphy, *Editor*

Vol 38: **Ethics, Humanism, and Medicine,** Marc D. Basson, *Editor*

Vol 39: **Neurochemistry and Clinical Neurology,** Leontino Battistin, George Hashim, and Abel Lajtha, *Editors*

Vol 40: **Biological Recognition and Assembly,** David S. Eisenberg, James A. Lake, and C. Fred Fox, *Editors*

Vol 41: **Tumor Cell Surfaces and Malignancy,** Richard O. Hynes and C. Fred Fox, *Editors*

Vol 42: **Membranes, Receptors, and the Immune Response: 80 Years After Ehrlich's Side Chain Theory,** Edward P. Cohen and Heinz Köhler, *Editors*

Vol 43: **Immunobiology of the Erythrocyte,** S. Gerald Sandler, Jacob Nusbacher, and Moses S. Schanfield, *Editors*

Vol 44: **Perinatal Medicine Today,** Bruce K. Young, *Editor*

Vol 45: **Mammalian Genetics and Cancer: The Jackson Laboratory Fiftieth Anniversary Symposium,** Elizabeth S. Russell, *Editor*

Vol 46: **Etiology of Cleft Lip and Cleft Palate,** Michael Melnick, David Bixler, and Edward D. Shields, *Editors*

Vol 47: **New Developments With Human and Veterinary Vaccines,** A. Mizrahi, I. Hertman, M.A. Klingberg, and A. Kohn, *Editors*

Vol 48: **Cloning of Human Tumor Stem Cells,** Sydney E. Salmon, *Editor*

Vol 49: **Myelin: Chemistry and Biology,** George A. Hashim, *Editor*